Communications Sector Protection and Homeland Security

ALSO IN THE CRITICAL INFRASTRUCTURE AND HOMELAND SECURITY SERIES

Chemical Infrastructure Protection and Homeland Security
Nuclear Infrastructure Protection and Homeland Security
Water Infrastructure Protection and Homeland Security
Energy Infrastructure Protection and Homeland Security
Food Supply Protection and Homeland Security
Dam Sector Protection and Homeland Security
Transportation Protection and Homeland Security
Government Facilities Protection and Homeland Security

Communications Sector Protection and Homeland Security

FRANK R. SPELLMAN

Homeland Security Studies

Bernan Press

Lanham • Boulder • New York • London

Published by Bernan Press
An imprint of The Rowman & Littlefield Publishing Group, Inc.
4501 Forbes Boulevard, Suite 200, Lanham, Maryland 20706
www.rowman.com
800-865-3457; info@bernan.com

6 Tinworth Street, London SE11 5AL

Library of Congress Cataloging-in-Publication Data

Names: Spellman, Frank R., author.
Title: Communications sector protection and homeland security : communications sector protection and homeland security / Frank R. Spellman.
Description: Lanham, Maryland : Bernan Press, an imprint of The Rowman & Littlefield Publishing Group, Inc., [2018] | Series: Homeland security series ; 10 | Includes bibliographical references and index.
Identifiers: LCCN 2018035516 (print) | LCCN 2018036923 (ebook) | ISBN 9781641433105 (electronic) | ISBN 9781641433099 (pbk. : alk. paper)
Subjects: LCSH: Telecommunication--Security measures--United States. | National security--United States. | Telecommunication policy--United States.
Classification: LCC TK5102.85 (ebook) | LCC TK5102.85 .S64 2018 (print) | DDC 384.068/4--dc23
LC record available at https://lccn.loc.gov/2018035516

∞™ The paper used in this publication meets the minimum requirements of American National Standard for Information Sciences—Permanence of Paper for Printed Library Materials, ANSI/NISO Z39.48-1992.

Printed in the United States of America

For Kathern Welsh

Contents

Acronyms and Abbreviations

APEC	Asia-Pacific Economic Cooperation
APM	Asset Prioritization Model
ATM	Asynchronous Transfer Mode
BIS	Bureau of Industry and Security
BSS	Broadcast Satellite Service
CATV	Cable Television
CERT	Computer Emergency Readiness Team
CII	Critical Infrastructure Information
CIKR	Critical Infrastructure and Key Resources
CIP	Critical Infrastructure Protection
CLEC	Competitive Local Exchange Carrier
CS/IA	Cybersecurity/Information Assurance
DCIP	Defense Critical Infrastructure Program
DHS	Department of Homeland Security
DIB	Defense Industrial Base
DoC	Department of Commerce
DoD	Department of Defense
Dodd	Department of Defense Directive
Dodin	Department of Defense Instruction
DoS	Department of State
DPAS	Defense Priorities and Allocations System
FEMA	Federal Emergency Management Agency

FOIA	Freedom of Information Act
GPS	Global Positioning System
HSPD	Homeland Security Presidential Directive
IA	Information Assurance
IP	Office of Infrastructure Protection
IT	Information Technology
IXC	Interexchange Carrier
LEC	Local Exchange Carrier
NCIP	National Critical Infrastructure Protection
NCS	National Communications System
PSTN	Public Switched Telephone Network
SSA	Sector Specific Agency
SSP	Sector Specific Plan
TSA	Transportation Security Administration

Preface

The tenth of a new, well-received and highly acclaimed series on critical infrastructure and homeland security, *Communications Sector Protection and Homeland Security* is an eye-opening account and an important reference source of a complex sector. This book was designed and written to serve and advise U.S. project designers, engineers, communications technicians, law enforcement and security specialists, managers, and superintendents and/or supervisors and responsible managers in charge of protecting the multifaceted nature of critical infrastructure in the United States. When I mentioned to a colleague of mine that I was in the process of writing the tenth volume of this series and I described the topic to be covered, she was surprised that I would embark on such a mind-numbing and difficult task. I remember that bewildered look on her face and her exact words when she asked: "How can you . . . or anyone . . . write about and describe the U.S. Communications Sector Infrastructure when it is so deep, tall, wide and all encompassing and almost indefinable in terms of total reach, scope, extent, and so forth?" Wow, I remember scratching my head and trying to answer her in an impressive way (not easy) and finally coming up with what for me was the ultimate answer. I replied, "If it exists, it can be written about and described." Then she scratched her head and smiled while shaking her head side to side (meaning no way, Jose) and then she just changed the subject.

Anyway, currently, the communications sector critical infrastructure sector includes underlying components that allow for the operations of all business, public safety organizations, and government. Basically, the Communications Sector, according to Presidential Policy Directive 21, is critical because it provides an "enabling function" across all critical infrastructures.

Well, this part of the description is plain to see in use and practice and easy to understand. Although it can deal with complex technology such as cloud computing, it does not take a "cloud" scientist to comprehend the complexity of this sector. And actually the complexity has evolved from the simplicity of simple transmission of communication via hand singles, flags, drums, horns and Morse code into a diverse, competitive, and interconnected industry using terrestrial, satellite, and wireless transmission systems employed in the Internet, cloud computing, and telecommunicating via various G-networks. The transmission of these services has become interconnected; satellite, wireless, and wireline providers depend on each other to carry and terminate their traffic and companies routinely share facilities and technology to ensure interoperability (DHS, 2017).

This book is organized to simplify and present in a logical and sequential manner a discussion of not only the elements comprising the electronic communications infrastructure in the United States but also many of the security measures employed to protect the various entities and equipment involved.

Let' s face today's reality. Those who want quick answers to complicated questions—to help employers and employees handle security threats—must be prepared to meet and deal with the threat of terrorism on a 24/7 basis. It is important to point out that this book does not discuss and focus on security concerns related to natural disasters; on the contrary, the focus here is on the security aspects in the design of systems that need to be able to deal robustly with possible sources of disruption, specifically from malicious acts; moreover, the focus includes the added dimension of preventing misuse and malicious behavior. In the post–September 11, 2001, world, the possibility of communications sector infrastructure terrorism—the malicious use of weapons and cyber intrusion to cause devastating damage to communications sector infrastructure and its associated subsectors along with—literally—its cascading effects—is very real. Thus, the need is clear and real and so are the format and guidelines presented in this text to improve protection and resilience of communications sector infrastructure.

This book describes the sector-subsector-wide process required to identify and prioritize assets, assess risk in the sector, implement protective programs and resilience strategies, and measure their effectives. This book (and the complete sixteen volumes (upgraded from the original fourteen volumes) of the critical infrastructure sector series) was written as a result of 9/11 to address these concerns. It is important to point that our communications sector infrastructure (as is the case with the other fifteen critical infrastructures) cannot be made absolutely immune to all possible intrusions/attacks; thus, it takes a concerted, well-thought-out effort to incorporate security upgrades in the retrofitting of existing systems and careful security planning for all new facility infrastructure components. These upgrades or design features need to address issues of monitoring, response, critical infrastructure redundancy, and recovery to minimize risk to the facility infrastructure. However, based on personal experience, none of these approaches is or can be effective unless communications sector staff members at all levels of the chain of command are cognizant of the threats.

Communications Sector Protection and Homeland Security presents common-sense methodologies in a straightforward, almost blunt manner. Why so blunt? At this particular time, when dealing with security of workers, family members, citizens and society in general—actually, with our very way of life—politically correct presentations on security might be the norm, might be expected, and might be demanded, but my view is that there is nothing normal or subtle about killing thousands of innocent people; mass murders certainly should not be expected; the right and need to communicate and the right to live in a free and safe environment is a reasonable demand.

This text is accessible to those who have no experience with or knowledge of the communications sector. If you work through the text systematically, you will gain an understanding of the challenge of domestic preparedness—that is, an immediate need for a heightened state of awareness of the present threat facing the communications sector members as potential terrorist targets. Moreover, you will gain knowledge of security principles and measures that can be implemented—adding a critical component not only to your professional knowledge but also giving you the tools needed to combat terrorism in the homeland—our homeland, both by outsiders and insiders.

One final word to readers: This book is written in the conversational and engaging and reader-friendly style that is the author's trademark. Why? Well, when demonstrating how one . . . or anyone . . . can write about the U.S. communications sector when it is so deep, tall, wide, and all-encompassing and almost indefinable in terms of total reach, scope, extent, and so forth, I never apologize for attempting to communicate.

Frank R. Spellman
Norfolk, Virginia

Prologue

No Calls in the Wild

Note: The following fictitious account is included here in order to set the stage for what follows in the text and to continue the sordid saga of the Williams family. The Williamses are homegrown radicalized terrorists (smart and brutal and not Muslim) who have done whatever they could do to bring about the death and destruction of innocents in the United States, with particular attention being paid to attacking, impacting and/or destroying sectors of critical infrastructure.

*****

I just hate those who text! Don't You?

Ok, let's have a one-way conversation . . . me to you . . . about this nightmarish practice and some of the other ills (there are so many to choose from) contaminating this country and society in general that need to be changed and/or eradicated . . . one way or the other . . . and I prefer the other, thank you very much.

All those phony electronic messages containing those even phonier alphabetic and numeric characters. Where has all the decency in communicating gone today? And have you noticed those frog-faced goonies driving down the Interstates weaving here and there and going off the road and into oncoming traffic and they just keep doing that until the drivers run into a car full of innocent folks (well, in my opinion, the word innocent is overused; I've never have met anyone who is innocent—know what I mean?) or a busload of holy rollers on their way to one of those phony churches where you are only welcome if you bring lots of money or you are a goodlooking person who can be readily seduced by those running church, and so forth. What a waste! Anyway, I think it is okay to kill . . . blood and guts here, there, and anywhere, but I have problem with those who unintentionally kill . . . like those texting fools.

Well, let's cut the baloney—you know, all that pretentious nonsense—and get down to brass tacks, so to speak. It certainly was not texting while driving and running into a vehicle full of so-called innocents that got me here in the slammer. No, sir. I am here because I am and I was unlucky. Here? Yes, it was because of communication towers and not because of texting that I am presently locked up in this cell in King County Lock-Up here in Seattle . . . my home town . . . where all the bleeding heart, ultra-brain-dead, liberal, dumb bunnies live. Oh, well, as I look around at the bars on the cell door and the three windowless steel walls painted sickening pea green, my gaze settles on those initials and short, pithy statements carved into those cold steel walls (I can't repeat those statements here, folks, because even I have some modesty . . . some . . . very little, if you must know). Most of the initials I am familiar with . . . many belong to my blood (blood means family for those of you who do not understand English and common and not bloody usage of the word blood) . . . I am part of that famous . . . excuse, me, infamous . . . family, the Williams. I am B. M. Williams VI . . . and proud of it. I know you have heard of us and our famous and patriotic exploits, which unfortunately end with so many of us inhabiting this isolation cell . . . sooner or later . . . usually sooner . . . before we have a chance to kill as many of you no-loads as possible. With us, failure is not an option . . . but unfortunately it is a common happening. But be advised, we are a large family and are just beginning our "Cleansing" of America and melting all you commie Snowflakes . . . and others who dare get in the way—remember, collateral damage is good . . . very good.

As I sit here in this pea-green coffin-like cell, I just hanker for the old days when we (people) used to talk to each other face to face, or on landline telephones, or Buddha forbid, write each other letters. It is true that at my young age, 24, thank you very much, I was born into the electronic age of personal computers, the Internet, e-mail and all those disgusting mobile devices. But even so, even as young as I am, I do remember talking on home telephones and writing notes to friends . . . yes, even I had a few friends, mostly patriots like me . . . mostly incarcerated like me.

Have you ever walked down the street, or in a store aisle and observed and overheard some goony speaking out load to no one in sight as he or she walks toward you? When I first witnessed such an occurrence I thought the other person talking out loud was loony . . . a nut case, wacko or one of them deranged snowflakes. But after a few months of witnessing the reoccurrence of the same behavior on the sidewalk, in grocery stores, and on buses, I came to realize that those crazies were actually talking to some other fool on the other end of one of those wireless communication networks. I ask you, where is the common decency of folks who have to speak out loud in movie theatres, stores or dressing rooms when other people have to hear their dumb comments to some unknown snowflake on the other end of the silly transmission? Disgusting, that is what it is. And . . . and, I really wonder if those jerks speaking out loud into their wireless phones are really speaking to anyone on the other end . . . after listening to what those snowflakes say out loud, it makes me wonder if there could actually be a true nutcase on the other end . . . or is the loud speaking lout trying to make all of us innocent listeners think that the person actually has someone to speak to . . . ? Just torques my jaws when people do not care about disturbing my peace and quiet. How about you? Bet it bothers you too . . . unless you are one of them loudmouthed snowflakes.

Ok, now all you poor souls out there in la-la land have been introduced to me and I know you are foaming at the bit to find out what it is that I have done that caused me to end up in this despicable, dreadful pea-green dungeon.

I am right, ain't I? You want to know all. You are foaming at the mouth in anticipation, aren't you? You want to know me better . . . you can't wait . . . I have you poor souls hooked, addicted to . . . well, I do not blame you one tiny bit . . . besides being extremely talented and ten times smarter than the average genius, I am rather appealing.

Well, ok, who am I to digress and keep you all waiting. Let's cut to the chase and get down to it . . . it being why I have joined many of my other bloods in occupying this horrible, inhumane lockup.

Once upon a time in la-la land . . .

It began after my first glance at one of the horrible analog 1G towers sticking up into the sky behind my house near the Interstate. What a horrible apparition to have to look up at each day as I wandered off to that commie private school I attended . . . for a while, at least (they eventually kicked me out for cutting off the girl's pigtails who was sitting in the desk in front of me) . . . anyway, that is a different story . . . back to those visible pollution sources piercing the sky. I think it was in 1999 (I was about seven or eight then) when they switched off those analogs (too much background noise and people listening in on conversations) . . . people are just sick, don't you agree? Anyway, technology advanced and they shifted to digital 2G, 3G, and 4G, and Buddha only knows what else! Whether it is 1G or any G, it is nothing more than visible pollution . . . disgusting, for sure.

Anyway, when I turned twenty-three, after years of maybe spiking old-growth trees so those commie lumber people could not cut them down without chopping themselves up . . . and maybe (I can't say for sure, the commie prosecutors are probably listening in) setting aflame all those gas-guzzlers in the those sleazebag car lots and doing other despicable things that, again, I can't comment on . . . otherwise those liberal pigs will prosecute me for things I might have accomplished in the past.

Anyway, my plan was conceived while I was in lockup for spitting on a liberal city council wench . . . can you believe that? My plan? Well, it was a brilliant plan, concept, and scenario or as that frog-faced DA called it, the No Call in the Wild caper. . . . Wow! What a title! Anyway, you need to understand two things that me and my blood share and it is our innate (probably infused) hatred for anything federal or state or local government that has anything to do with enforcing those stupid laws those commies put into effect . . . and the other thing me and my blood share: Our genius, of course. Einstein, Newton, Aristotle, Stonewall Jackson, Patton and da Vinci, well, they were pretty smart, for sure . . . well, the Williams are just a whole lot smarter . . . our only problem is bad luck . . . lots of bad luck and, when employed, incompetent coconspirators.

Anyway, I said it is time to cut to the chase . . . so let me tell you about the so-called No Call from the Wild caper . . . holding you all in this suspense is killing me, too . . . ha, ha . . . not really.

I wanted to follow family tradition and example to destroy as much property and kill off as many despicable humans as possible. Now, this admirable goal is obtainable but, at the same time, it is difficult to achieve. Ever since my college days (entered college at fourteen) as an undergrad, grad, and post-doc student and researcher, I had been involved in family actions against this or that . . . but, again, I can't relate those details because I have to get through this mess and get out of prison and continue with my future grand plans.

A few years ago, I came up with a plan to destroy communication towers. I was and am particularly interested in cellular and microwave towers . . . those radio and television towers are of no interest to me. The towers themselves are relatively easy targets even though they are so widespread and there are so many of them . . . my research in 2016 enabled me to discover that at that time there were almost 220,000 cell phone towers throughout the United States . . . and I found out that the maximum range of a cell phone tower is as much as 22 miles. Also, according to the Federal Communications Commission's 200 Antenna Structure Registry, tower construction has been growing by an estimated 6 to 8 percent annually . . . that's a lot of towers . . . too many!

Again, my view of cell towers is that they are instruments of visual pollution of the worst sort. Ah, but there is more to my hateful and disgusting views of these human-constructed monstrosities . . . yes, more. How about all the birds they kill? Cell phone towers kill something like 4–5 million birds per year! Especially vulnerable to striking one of these towers and dying as a result are some 350 species of night-migrating birds. Why, you ask? Simply, many of the towers are poorly lighted (off-lighted instead of down-lighted onto the tower itself) or not lighted at all . . . especially those towers located in remote locations. It is the towers in remote places that are of interest to me . . . for obvious reasons.

You ask, for obvious reasons? Yes, my targets had to be in fairly remote places. I had to whittle down to, so to speak, a specific number of targets . . . ones that were doable, destroyable. I came up with a brilliant plan to attack and destroy a large number of remote towers.

Because I have already been convicted for my actions, I figure nothing is to be lost by telling you what my plans were . . . about the so-called No Call in the Wild caper . . . I will leave some unknown factors out for obvious reasons. There were a few things I knew based on experience and research. First, I knew I had to work alone. I am not a lone wolf . . . more like a lone Utah raptor, thank you very much! Anyway, based on my family's past experience with partners in crime (so to speak, spiking trees, trashing car lots, blowing up things and killing people for good reason is never a crime . . . not in my family, thank you very much!) they found out the hard way that adding people to a plan is liking adding weak links to an extensive chain. Never work as a team; instead, work alone. For example, when you rob a bank, if you work alone, all you have to worry about is the bank teller and the video cameras . . . and if you have to, you can destroy both . . . but if you get away scot free then the only one who knows what you accomplished is you. I do not think you will rat yourself out unless, that is, you are one of them commie froggers . . . you know the ones I mean . . . the snowflakes on drugs, alcohol, or who are walking the earth with a head full of loose screws . . . or they simply have big mouths . . . and most do have big mouths.

Anyway, secondly, never abide by or stick to the eight signs of terrorism that the commie homeland security and other officials look for. Today, after 9/11, more people are paying attention to those eight signs . . . and they can give you up before you even destroy something or humans; not good. What are the eight signs? Gee, thought you would never ask. Actually, the eight signs are published and described by that Big Brother agency known as FEMA . . . I did my homework, important. You remember FEMA don't you? Well, just think Hurricane Katrina . . . ya, you got it . . . real mess by an incompetent, political agency. Ok, here we go . . . I will bulletize them . . . I just love bullets (bombs, A-bombs, and firebombs included):

- Surveillance
- Elicitation
- Testing security
- Funding
- Acquiring supplies
- Suspicious persons
- Rehearsal phase
- Deployment

Okay, humor me and let me go down the list and tell you what I did and did not do. Remember, the Williams family members are geniuses . . . so we do know what to do and not do . . . well, sort of. With regard to surveillance, I already knew my targets from my hiking and nature outings and a few tree-spiking activities. My target area was all the towers . . . more than a dozen on the Olympic Peninsula in Washington State . . . the place is a paradise but I have no time to describe it all here. From Neah Bay located near the farthest point of land that sticks out at the terminus of the north-west point of Washington State into the Pacific to almost the eastern point of the Olympic Peninsula on the east side of the city of Port Angeles. You know when you are backpacking without a 12-gauge or Dirty Harry Special in your hands people seem to ignore you . . . so when I hiked and did my surveillance of the fourteen towers along the route just described no one gave me much notice. Even my red-headed beauty and knockout body did not get their attention . . . and that was good.

Now, in regard to elicitation (what a stupid word, why not just say entice information from dumb-dumb clowns), it was not necessary for me to spend any time in this area because of the information that is readily available via the Internet. And, let me ask you this, how does one entice information from people who do not exist? My point is that it is rare to see any worker at or near one of the towers unless there is a malfunction or installation of some new technological device or devices. And if they did show up, I could talk a frog out of her warts . . . that is the advantage of total beauty in all respects; it opens many mouths . . . and them snowflakes are eager to spill the beans, so to speak . . . and I speak from observation and experience.

Testing security is another key step in terrorist planning . . . no kidding, Jose and Maria. But . . . however . . .there is no security that I knew of at those tower locations . . . they are remote and not guarded in any way I was aware of.

It is expensive to accomplish a terrorist mission . . . but . . . without going into detail, funding has never ever been an issue for the Williams family . . . enough said.

I had all the supplies I needed for this caper. Well, I had a load of military grade explosives and rolls of twine to attach the explosives where they needed to be secured. Enough said.

Suspicious persons or impersonation? Wow, really? Simply, this did not apply to me because there is nothing suspicious about me and I did not need to interface with anyone. And, who would I impersonate? No one on Earth that I can think of is suitable to impersonate.

In my case there was no rehearsal needed. I knew the targets and how to reach them and that they were mostly in secluded areas off the highway where no one was around and only hikers in the woods might notice me. The exact and precise plan to blow up the towers in each of their locations was clearly imprinted in my mind. I knew what had to be done and how to accomplish it.

It was the deployment aspect that turned out to be my Achilles's heel . . . my downfall. What was my downfall, you ask? Well, I am getting to that . . . patience, please . . . let me catch my breath and composure . . . failure is hard to swallow.

Okay, gulp, gulp, gulp . . . what happened is this. Even with all my brilliance and attention to detail and my unmatched ability to get the job done, I ran into the same problem my other family members encountered: Bad luck.

I began my caper near Neah Bay and planned to do each tower in succession. I planned on attaching enough explosives to each tower to bring them down and destroy their usefulness. I intended to install the explosive charges on each tower and wait until they were all in place before detonating them remotely. I intended to accomplish one-by-one detonations of each tower while driving down Washington Highway 112 back to Neah Bay and then to turn around and head down the road toward Port Angeles.

Well, you know what they say about the best laid plans . . . and so forth . . . my plan failed right after I had placed my first explosive charge at the first targeted tower while I drove to tower number 2 . . . it was a misty early morning with little traffic, either way, but as I drove down the slick highway a woman driving the opposite direction, while texting, of course, came across the centerline and struck me near the rear driver's side door, spun me out, wrecked the vehicle and to my good luck the explosives in the trunk did not detonate . . . otherwise I would not be here talking to you in this sickening lockup. The bad luck was twofold: the wreck itself, of course, did not allow me to accomplish my mission; also, just as I got out the car in a daze of sorts, I saw those sickening flashing lights of the commie State Patrol as they arrived at the wreck site.

Well, needless to say, it did not take long for the male trooper to discover my deadly and illegal cargo in my wrecked car. The trooper's partner, a female that was built like a wrestler and looked like the offspring of Freddie the Kruger grabbed me, slapped the crap out of me, threw me to the ground and handcuffed me and basically picked me up and threw me into the back of the squad car (normally, I just love take-charge women, don't you?).

Months later I had my trial and after that two unsuccessful appeals; I was totally shocked, to say the least, that someone like me who was as guilty as they come was actually found guilty; hard to believe in this day and age of liberal juries and judges. Oh, well, I did not kill anyone so I received an eight-year sentence and with good behavior will be out to accomplish future capers in about two years . . . or so. To restate a famous line: I will be back!

But in the meantime, right now I hear one of those terrible screws speaking to someone in the cellblock hallway outside my cell. I looked to find out who they were and was surprised to see and hear one screw walking down the hall talking to someone (supposedly) on the other end of a wireless connection. Ironic, you know, they called my caper No Call in the Wild . . . and I just heard and witnessed a Call from the Wild . . . silly? Maybe. So, until next time . . . in the future . . . whenever I accomplish my next mission, please do not insult me by calling me . . . write me a note of appreciation instead.

1

Introduction

Fear is the emotion that makes us blind.

—Stephen King

Never underestimate the time, expense, blood, sweat, and effort a terrorist will apply to compromise the security of any industrial facility.

In America, there are plenty of hate groups that claim peace and brotherhood, but when their actions are responsible for death and destruction, they are identified for what they really are.

—Lindsey, 2001

. . . passion and party blind our eyes, and the light which experience gives is a lantern on the stern, which shines only on the waves behind us.

—Samuel Coleridge

WE FEEL IT BECAUSE OF A NEED; IT IS THE NEED TO COMMUNICATE

Have you ever paced back and forth across your living room floor, then stopped and looked out a window, trying to shake the grips of depression? And then watch the raindrops strike against the window pane. Off in the distance, a foghorn sounds its deep solitary tone, forlorn. You hear the old grandfather clock chiming the hour. You count the chimes: three . . . six . . . nine. They stop at twelve. What is wrong? Why the depression? Why that feeling of total emptiness? Why the welling of tears? What is wrong? The cause of your severe depression could be any number of things; there are probably too many causes to list. What is wrong? Again, we could list a number of reasons. But for our purposes in this text and in this instance, and for the moment, we will pick just one: loneliness.

What is loneliness? Well, it was described above, at least to a degree. Those who are lonely know what loneliness really is. Loneliness is probably best described as one's unfulfilled desire for another person's company. Loneliness makes one feel disconnected and isolated from everyone else. So, how do we cure loneliness? Is there a pill we can take? Is there some magic potion available to curb loneliness? Is there some diversion to curb loneliness? All of these methods have been tried; we know of instances where these actions actually occur. But the bottom line, the actual cure for loneliness is companionship; it is probably the best medicine, potion, or diversion for loneliness.

If companionship is the best solution to curbing one's loneliness, the question is how do we accomplish this when it is almost impossible for people to be together at all times? Simply, communication is the solution to isolation and loneliness. When we can pick up a phone and talk with someone we know or even a counselor or someone in a like capacity, we are communicating with someone else. Years ago, when I explained this personal opinion to a large class of college undergrads in one of my environmental health classes, one of my students raised her hand and said, "Professor, what if someone calls someone on the phone and the answer is not a real person but instead is a recording?" I smiled and at first wanted to answer, "Ah, great question, grasshopper." Instead I replied, "I think the sound of another human's voice . . . even an artificial human voice . . . on the other end of the line indicates, somewhat, to a degree, to the caller that he or she is not alone. We need it . . . communication . . . it is a human need and conducted properly curbs loneliness. But be advised, modern-day communication accomplishes much more and is needed for other important purposes."

So, what are the other purposes of communication? Actually communication has more of an objective than a purpose. In general terms we communicate in order

- to provide or give information.
- to persuade.
- to seek information.
- to express emotions like joy or sorrow, fear, satisfaction or disappointment.

Okay, to this point we have pointed out a few of the major reasons why we communicate. However, this book is more concerned with communications technology—which has significantly changed the way people communicate—and, more to the point, protection of the communication infrastructure.

As mentioned, communication infrastructure technology has significantly changed, and society has accepted and embraced the increased role technology plays in everyday life. Originally, the telephone replaced the telegraph. Now cell phones, email and the Internet top the list of preferred communication methods. It is the protection of the infrastructure that supports these communication methods that this book is about.

SECTOR OVERVIEW

The primary entity responsible for protecting communication sector infrastructure and assets is the private sector; it is the owner and operator of the majority of communications infrastructure. In conjunction with the federal government, the private sector is able to predict, anticipate, and respond to sector outages and understand how they might affect the ability of the national leadership to communicate during times of crisis, impact the operations of other sectors, and affect respond and recovery efforts.

The communications sector is closely linked to other sectors. Specifically, communications systems are the backbone for much of the critical infrastructure within the United States, and many of the other infrastructure components are completely dependent on communications systems to perform their missions. The communications sector provides the basis for information exchange for all other sectors including voice, data, video, and internet connectivity. As such, communications systems serve with other key national security and emergency preparedness resources and are an important component of our overall national critical infrastructure. It is therefore necessary to recognize the importance that communications has for national security policy and issues of homeland security. At the same time, it is important to recognize the key factors upon which the communications infrastructure is dependent. The focus here is on the interdependencies between the communications sector and the national utility infrastructure. Moreover, the communications sector is closely linked to other critical sectors (each of which is covered or are in production in separate editions of this series), including (DHS, 2017):

- The *Energy Sector*, which provides power to run cellular towers, central offices, and other critical communications facilities and also relies on communications to aid in monitoring and controlling the delivery of electricity.
- The *Information Technology Sector*, which provides critical control systems and services, physical architecture, and internet infrastructure, and also relies on communications to deliver and distribute applications and services.
- The *Financial Services Sector*, which relies on communications for the transmission of transactions and operations of financial markets.
- The *Emergency Services Sector*, which depends on communications for direction resources, coordinating response, operating public alert and warning systems, and receiving emergency 9-1-1 calls.
- The *Transportation Systems Sector*, which provides the diesel fuel needed to power backup generators and relies on communications to monitor and control the flow of ground, sea, and air traffic.

At this point you have probably surmised that the one way of investigating the interdependence of the communications architecture is to look at it from a critical infrastructure perspective. The National Infrastructure Protection Plan (NIPP) is intended to meet the requirements set forth by the president in Presidential Directive 7 (HSPD-7), Critical Infrastructure Identification, Prioritization, and Protection, and is an overarching approach for integrating the nation's Critical Infrastructure and Key Resources (CIKR) protection initiatives.

One of the critical sectors within the CIKR is the Communications Sector. In addition, the Energy Sector (covered in detail in Volume 2 of this series) is comprised of the oil, gas, and electric power production, refining, storage and distribution facilities, and the Water Sector (covered in detail in Volume 1 of this series) includes all those facilities for fresh water and wastewater management. These two sectors are administered by national or local public service commissioners (DHS, 2017).

The national protection plan highlights those sectors that are critical for national security and emergency preparedness and delineates measures to protect these resources. In the same way, the Communications Sector-Specific Plan details factors that are important for the protection of the communications sector. On the other hand, the National Association of Regulatory Utility Commissioners (NARUC) has published a set of Critical Infrastructure Technical Briefs that "identify key strategies for our consideration as we meet ongoing challenges within each of the electricity, natural gas, water, and telecommunications sectors." Figure 1.1, based on the diagram provided in NARLUC *Issue Paper on Critical Infrastructure Protection,* shows many of the interdependencies for the utility "sector" (FCC, 2017), partially illustrates the extent of the interdependencies between the utilities (electric power, oil and gas, water), and other sectors including communications, transportation, banking and finance, and government services.

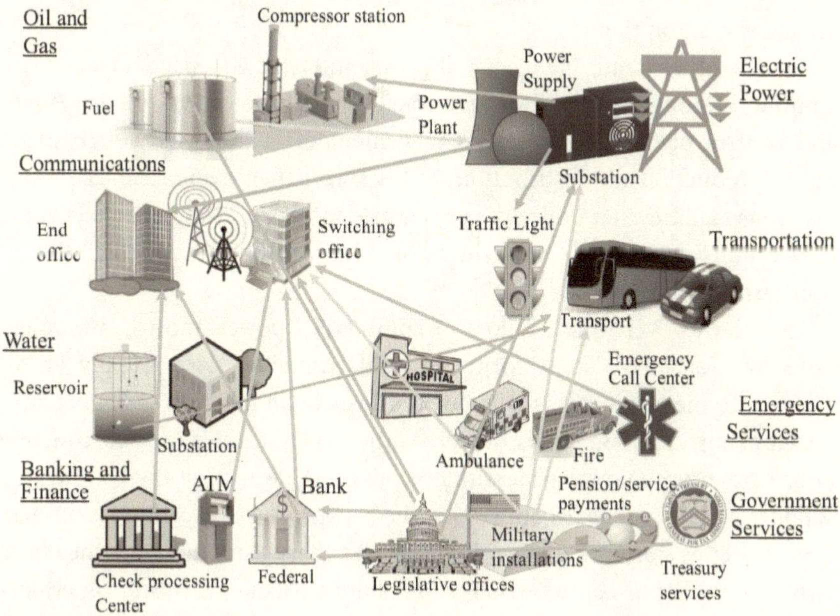

FIGURE 1.1

Partial illustration of the interdependencies between utilities and other critical infrastructures.

Source: Adapted from FCC (2017) and illustrated by F. R. Spellman and Kathern Welsh.

It is clear from Figure 1.1 that there is a great deal of interdependency between the Communications Sector and a number of the functionaries within the utility community. Figure 1.1 illustrates that almost all of the utilities have critical requirements for communications of some form. Alternatively, the communications community has a number of instances where they are dependent on the utilities. The remainder or this section will highlight some of these issues.

Without a doubt, the dominant dependency for the communications sector is electrical power. Every node in the communications architecture—whether it is a switching center, radio relay site, cell site, other remote site, or any other facility—relies on electrical power for its operation. Electricity powers the communications systems equipment, the central control/management/operations systems, and even the environmental control systems surrounding the communications equipment. In most cases, the power to run the communications infrastructure is provided on a continual basis by the commercial power industry. As such, there is a direct and critical link between the electrical power networks and the communications networks that are dependent on them.

On the other hand, the electrical power industry is dependent upon the communications providers for inter-facility communications, management and control of operations, and management of facilities. Although many power production and distribution facilities have their own organic communications facilities that utilize the power grid infrastructure, nevertheless, the power industry is still reliant on the commercial communications facilities for transferring information.

In order to overcome any loss of commercial power, most communications locations have some form of uninterruptable power supply (UPS) inherent to the facility. This UPS power is usually in the form of battery support for short duration power outages while larger facilities will subsequently utilize emergency power generators to provide power until the commercial supplier is back on line. In both cases, the backup power capability is limited. First, batteries store energy and do not create energy and therefore they have a short-term life of approximately a few hours. Second, the local generators have a limited supply of fuel and must be resupplied depending on how long the generators must continue to operate (FCC, 2017).

During the 2003 Northeast region blackout and Hurricane Katrina in 2005 and other recent large-scale power-outages the strong interdependencies of the power and communications sectors were highlighted. In response to these outages and the resulting studies done by the National Security Telecommunications Advisory Committee (NSTAC), in 2007 the National Communications System Committee of Principals (COP) established the Communications Dependency on Electrical Power Working Group (CDEPWG). The report of this working group found a number of conclusions and recommendations for the communications sector to address long-term power outages and their impacts on the communications sector (FCC, 2017).

The requirements for locally generated power suggest another dependency for the communications sector. Where it is necessary to generate power independently; the generators rely on some source of energy (water, fuel, etc.) to run the generators and this is usually a fossil fuel of some form, usually diesel fuel. Thus, the reliance on the energy sector is critical for communications providers when there is a loss of commercial power and for as long as they must run their generators. This is not the rule, though, and usually the communications providers have limited on-hand supplies of fuel and rely on support agreements with local fuel and transportation providers to be able to support them in times of disaster, outages, or other emergencies. Closely related to fuel for generators is the requirement in many cases for natural gas or other fuels for heating purposes. Note that most communications facilities require environmental control to maintain nominal conditions for equipment to operate, this may require either water for cooking or gas for heating.

Keep in mind that the dependency on water resources as a source for energy to generate power is most likely a small factor when compared to the uses of water for other purposes. Access to water and the ability to handle wastewater can be critical dependencies for communications providers. In most cases, the environmental control systems for communications facilities rely on fresh water for air conditioning and other environmental services. The availability of fresh water is therefore extremely important for facility owners and operators. Similarly, the ability to handle wastewater from chillers or coolers is also important. Both the supply of fresh water and the handling of wastewater are essential for most large-scale communications facilities.

THE BOTTOM LINE

In addition to the critical dependencies for the communications sector mentioned to this point, it is important to note that physical security of communications sector facilities (coved later) is vital, but, actually, the focus of security of this sector is a different kind. Much of the communications infrastructure is vulnerable to cyber attack from either inside or outside the network. Cybersecurity is addressed later. Also, communications infrastructure and the utilities are extremely dependent on the information technology sector (covered in detail in another of the volumes in this series).

The bottom line is that from this discussion, it is clear that any number of interdependencies could threaten one of the key components of our critical national infrastructure, the communications sector. These factors are the "realities" of the communications sector and go beyond just the core precepts of good communications systems development and design. Clearly, there is a great deal of interdependence with other outside factors that could threaten communications and other key resources. Without the active participation of each of these sectors, the key and essential facilities of the communications industry could be very vulnerable, a fact that is not acceptable for the protection of our national interests (FCC, 2017).

REFERENCES AND RECOMMENDED READING

DHS (2017). *Communications Sector.* Washington, DC: Department of Homeland Security. Accessed 04/06/2017. https://www.dhs.gov/commuincaitns-sector.

FCC (2017). Public Safety Tech Topic #19—Communications Interdependencies. Accessed 8 April 2017. https://www.fcc.gov/help/public-safety-tech-topic-19-communiicaitons-interdependices.

Haimes, Y. Y. (2004). *Risk Modeling, Assessment, and Management*, 2nd Edition. New York: John Wiley & Sons, 699.

Henry, K., (2002). New face of security. *Gov. Security*, Apr., pp. 30–37.

NSHS (2006). *National Strategy for Homeland Security.* Accessed 05/13/06. www.whitehouse/homeland.

Sauter, M. A., and Carafano, J. J. (2005). *Homeland Security: A Complete Guide to Understanding, Preventing, and Surviving Terrorism.* New York: McGraw-Hill.

Spellman, F. R. (1997). *A Guide to Compliance for Process Safety Management/Risk Management Planning (PSM/RMP).* Lancaster, PA: Technomic Publishing Company.

2

Terrorist Attacks Against Americans

America is no longer protected by vast oceans. We are protected from attack only by vigorous action abroad, and increased vigilance at home.

—President George W. Bush, 2002

INTRODUCTION

It was not that far in America's distant past that we dealt mostly with the ramifications of natural disasters. Although never welcomed, these natural occurrences were expected, planned for, guarded against, and recovered from. In regard to disasters, they were our primary concern. Of course, there have always been the human-made or human-caused disasters that we also had to contend with. The sinking of the *Titanic* prompted new safety regulations for passenger ships; a fire at New York City's Triangle Shirtwaist Factory in 1911 caused the deaths of 146 young women and was the catalyst for the development of workplace safety and fire regulations. However, the ultimate wake-up call for Americans came in the form of the events that we now summarize as "9/11" or as some characterize as Cantor Fitzgerald 9/11 (because of the 658 employees of this business firm that were killed in the World Trade Center on 9/11/2001). Call it what you will; one fact remains, however: on September 11, 2001, terrorists struck at the heart of America—on American soil in a way that is unforgettable. Airplanes filled with people and fuel were turned into guided missiles of death and destruction.

What the 9/11 terrorists provided us with was a wake-up call. It was a wake-up call that alerted us to a new type of hateful venom—coldly delivered by groups or individuals to kill massive numbers of people, cause substantial property damage, and affect economic stability. Most importantly, for the public and emergency responders the events of 9/11 brought to the forefront knowledge and awareness that just about anything unthinkable is possible. The bottom line: security of the homeland is vital to all of us.

As noted by that great philosopher (Bob Dylan, of course) in the song, "The Times They Are A-Changin'," recent terrorist events have made this well-worn statement certainly the case. For example, with the manifestation of the Islamic State of Iraq and Syria (ISIS), also known as the Islamic State of Iraq and the Levant (ISIL) and by is Arabic language acronym Daesh, and other terrorist groups, including inside and outside fanatics, there are many out there who would kill us all. It is important to point out with the advent of insider threats that the message of this book is the same as the other books in this series where the focus has been shifted to the Lone Wolf (or as Williams VI stated in the prologue, lone Utah Raptor) terrorist. The fact is there are homegrown terrorists who live among us, plotting, quivering in the glory of mass murder, blood and guts and beheadings, destruction and the attendant glory of making the headlines as martyrs in the sickest sense possible. Again, remember, these are people who live among us, who shop in our stores, who use the benefits of our free society to grow their hate and to enhance their total disgust for all those things that we value in life.

Before 9/11 terrorism was not previously seen as a significant threat to the US. However, while the United States has not experienced continuous 9/11-type terror events in the past, the current climate suggests that the United States is at significant risk of further terror activity in the future. Forecasting the future is impossible, but we can look at the past and learn from it. Table 2.1 summarizes past acts of sabotage/terrorism carried out in the US, or against US citizens overseas.

Table 2.1. Terror Attack Summaries (within US or against Americans abroad)[1]

Date/Location	*Summary of Attacks*
Sept. 16, 1920 New York City	Bomb exploded in New York City's Wall Street area, killing 40 and injuring hundreds. Perpetrators fled country and were never apprehended.
1951–1956 New York City	Between 1951 and 1956, former Consolidated Edison employee set off series of bombs at New York City landmarks, including Grand Central Station and Radio City Music Hall. No deaths. Bomber was later caught and committed to state mental institution.
Jan. 24, 1975 New York City	Bomb set off in historic Fraunces Tavern killed 4 and injured more than 50 people. Puerto Rican nationals group (FALN) claimed responsibility, and police tied 13 other bombings to the group.
Nov. 4, 1979 Tehran, Iran	Iranian radical students seized the U.S. embassy, taking 66 hostages. 14 were released. The remaining 52 were freed after 444 days on the day of President Reagan's inauguration.
1982–1991 Lebanon	Thirty US and other Western hostages kidnapped in Lebanon by Hezbollah. Some were killed, some died in captivity, and some were eventually released. Terry Anderson was held for 2,454 days.
April 18, 1983 Lebanon	US embassy destroyed in suicide car-bomb attack; 63 dead, including 17 Americans. The Islamic Jihad claimed responsibility.
Oct. 23, 1983 Lebanon	Shiite suicide bombers exploded truck near US military barracks at Beirut airport, killing 241 marines. Minutes later a second bomb killed 58 French paratroopers in their barracks in West Beirut.
Dec. 12, 1983 Kuwait City, Kuwait	Shiite truck bombers attacked the US embassy and other targets, killing 5 and injuring 80.
Sept. 20, 1984 East Beirut, Lebanon	Truck bomb exploded outside the US embassy annex, killing 14, including 2 US military.
Dec. 3, 1984 Beirut, Lebanon	Kuwait Airways Flight 221, from Kuwait to Pakistan hijacked and diverted to Tehran. 2 Americans killed.
April 12, 1985 Madrid, Spain	Bombing at restaurant frequented by US soldiers, killed 18 Spaniards and injured 82.
June 14, 1985 Beirut, Lebanon	TWA Flight 847 en route from Athens to Rome hijacked to Beirut by Hezbollah terrorists and held for 17 days. A US Navy diver executed.
Oct. 7, 1985 Mediterranean Sea	Gunmen attacked Italian cruise ship, *Achille Lauro*. One US tourist killed. Hijacking linked to Libya.
Dec. 18, 1985 Rome, Italy, and Vienna, Austria	Airports in Rome and Vienna were bombed, killing 20 people, 5 of whom were Americans. Bombing linked to Libya.
April 2, 1986 Athens, Greece	A bomb exploded aboard TWA flight 840 en route from Rome to Athens, killing 4 Americans and injuring 9.
April 5, 1986 West Berlin, Germany	Libyans bombed a disco frequented by US servicemen, killing 2 and injuring hundreds.

(*Continued*)

Table 2.1. (Continued)

Date/Location	*Summary of Attacks*
Dec. 21, 1988 Lockerbie, Scotland	N.Y.-bound Pan-Am Boeing 747 exploded in flight from a terrorist bomb and crashed into Scottish village, killing all 259 aboard and 11 on ground.
Feb. 26, 1993 New York City	World Trade Center bombing in New York City. Six deaths and 1,000 injuries. Intended plan of total structural collapse did not occur. Several members of Middle Eastern extremist organizations convicted of roles in the bombing.
Dec. 7, 1993 New York City	Immigrant opened fire on commuters aboard Long Island Railroad commuter train, killing six and wounding 19. Gunman convicted of shootings dubbed "the Long Island Railroad Massacre."
April 19, 1995 Oklahoma City	Car bomb exploded outside federal office building, collapsing walls and floors. 168 people were killed, including 19 children and 1 person who died in rescue effort. Over 220 buildings sustained damage. Timothy McVeigh and Terry Nichols later convicted in the anti-government plot to avenge the Branch Davidian standoff in Waco, Texas, exactly 2 years earlier.
Nov. 13, 1995 Riyadh, Saudi Arabia	Car bomb exploded at US military headquarters, killing 5 US military servicemen.
Oct. 9, 1995 Hyder, Arizona	12-car Amtrak passenger train traveling from New Orleans to Los Angeles carrying 248 people derailed killing one and seriously injured 12. Investigations indicated that tracks had been deliberately tampered with. Incident classified as deliberate act of sabotage. Ongoing investigation in progress.
June 25, 1996 Dhahran, Saudi Arabia	Truck bomb exploded outside Khobar Towers military complex, killing 19 American servicemen and injuring hundreds of others. 13 Saudis and a Lebanese, all alleged members of Islamic militant group Hezbollah, were indicted on charges relating to the attack in June 2001.
July 27, 1996 Atlanta, Georgia	Bomb exploded during concert at Summer Olympics, killing one and injuring 118.
October 1997 San Francisco, CA	An outage at electrical substation in San Francisco affected 126,000 customers. Pacific gas and Electric said at the time that someone might have deliberately manipulated equipment at the substation to break circuits.
Aug. 7, 1998 Nairobi, Kenya, and Dar es Salaam, Tanzania	Truck bombs exploded almost simultaneously near 2 US embassies, killing 224 (213 in Kenya and 11 in Tanzania) and injuring about 4,500. 4 men connected with al-Qaeda, 2 of whom had received training at al-Qaeda camps inside Afghanistan, were convicted of the killings in May 2001 and later sentenced to life in prison. A federal grand jury had indicted 22 men in connection with the attacks, including Saudi dissident Osama bin Laden, who remained at large.
May 2000 Worldwide impact	Computer hacker in Philippines unleashed "Love Bug" computer virus, overloading corporate and government e-mail systems in many countries. Estimated $10–15 billion in damage. Charges against hacker eventually dropped in Philippines because of inadequate laws against computer hacking.

Date/Location	Summary of Attacks
July 25, 2000 Detroit, Michigan	Saboteur with knowledge of Detroit lighting system ripped wiring out of 100 streetlights in downtown Detroit, leaving live wires exposed and forcing city officials to shut off power to more than 600 downtown streetlights to ensure public safety. Damages estimated at $26,000. One person was arrested and charged.
Sept. 11, 2001 New York City, Washington, DC, and Pennsylvania	Terrorist cells of Middle Eastern origination hijacked four commercial jet airliners. Two aircraft used against World Trade Center Towers in NYC, third against Pentagon. Fourth aircraft brought down by passenger intervention. Approximately 3,000 killed.
September 2001 US	US Postal system used as delivery vehicle for anthrax spores contained in letters. Numerous news organizations and government agencies affected across US and in some overseas locations with strong US affiliations. Several Killed.
Oct. 7, 2001 Alaska	Intoxicated man shoots at the Alaska pipeline, which results in a 7,000 barrel crude oil spill, and the pipeline being shut for several days.
May 12, 2003 Riyadh, Saudi Arabia	Suicide bombers kill 34, including 8 Americans at housing compounds for Westerners. Al Qaeda suspected.
May 29–31, 2004 Riyadh, Saudi Arabia	Terrorists attack the offices of a Saudi oil company in Khobar, Saudi Arabia, take foreign oil workers hostage in a nearby residential compound, leaving 22 people dead, including one American.
June 11–19, 2004 Riyadh, Saudi Arabia	Terrorists kidnap and execute Paul Johnson, Jr,, an American, in Riyadh, Saudi Arabia. 2 other Americans and BBC cameraman killed in gun attacks.
Dec. 6, 2004 Jeddah, Saudi Arabia	Terrorists storm the U.S. consulate, killing 5 consulate employees. 4 terrorists were killed by Saudi security.
Nov. 9, Amman, Jordan	Suicide bombers hit 3 American hotels. Radisson, Grand Hyatt, and Days Inn, in Amman, Jordan, killing 57. Al Qaeda claimed responsibility.
Sept. 13, 2006 Damascus, Syria	An attack by four gunman on the American embassy is foiled.
Jan. 12, 2007 Athens, Greece	The U.S. embassy is fired on by an anti-tank missile, causing damage but no injuries.
Dec. 11, 2007 Algeria	More than 60 people are killed, including 11 United Nations staff members, when Al Qaeda terrorists detonate two car bombs near Algeria's Constitutional Council and the United Nations offices.
May 26, 2008 Iraq	A suicide bomber on a motorcycle kills six U.S. soldiers and wounds 18 others in Tarmiya.
June 24, 2008 Iraq	A suicide bomber kills at least 20 people, including three U.S. Marines, at a meeting between sheiks and Americans in Karmah, a town west of Baghdad.
June 12, 2008 Afghanistan	Four American servicemen are killed when a roadside bomb explodes near a U.S. military vehicle in Farah Province.
July 13, 2008 Afghanistan	Nine U.S. soldiers and at least 15 NATO troops die when Taliban militants boldly attack an American base in Kunar Province, which borders Pakistan. It's the most deadly attack on U.S. troops in three years.

(*Continued*)

Table 2.1. (Continued)

Date/Location	*Summary of Attacks*
Aug. 18 and 19, 2008 Afghanistan	As many as 15 suicide bombers backed by about 30 militants attack a U.S. military base, Camp Salerno, in Bamiyan. Fighting between U.S. troops and members of the Taliban rages overnight. No U.S. troops are killed.
Sept. 16, 2008 Yemen	A car bomb and a rocket strike the U.S. embassy in Yemen as staff arrived to work, killing 16 people, including 4 civilians. At least 25 suspected al-Qaeda militants are arrested for the attack.
Nov. 26, 2008 India	In a series of attacks on several of Mubai's landmarks and commercial hubs that are popular with Americans and other foreign tourists, including at least two five-star hotels, a hospital, a train station, and a cinema, about 300 people are wounded and nearly 190 people die, including at least 5 Americans.
Feb. 9, 2009 Iraq	A suicide bomber kills four American soldiers and their Iraqi translator near a police checkpoint.
April 10, 2009 Iraq	A suicide attack kills five American soldiers and two Iraqi policemen.
June 1, 2009 Little Rock, Arkansas	Abdulhakim Muhammed, a Muslim convert from Memphis, Tennessee, is charged with shooting two soldiers outside a military recruiting center. One is killed and the other wounded. In a January 2010 letter to the judge hearing his case. Muhammed asked to change his pleas from not guilty to guilty, claimed ties to al-Qaeda, and called the shooting a jihad attack to fight those who wage war on Islam and Muslims.
Dec. 25, 2009	A Nigerian man on a flight from Amsterdam to Detroit attempted to ignite an explosive device hidden in his underwear. The explosive device that failed to detonate was a mixture of powder and liquid that did not alert security personnel in the airport. The alleged bomber, Umar Farouk Abdulmutallab, told officials later that he was directed by the terrorist group Al Qaeda. The suspect was already on the government's watch list when he attempted the bombing; his father, a respected Nigerian banker, had told the U.S. government that he was worried about his son's increased extremism.
Dec. 30, 2009 Iraq	A suicide bomber kills eight American civilians, seven of them CIA agents, at a base in Afghanistan. It's the deadliest attack on the agency since 9/11. The attacker is reportedly a double agent from Jordan who was acting on behalf of al-Qaeda.
May 1, 2010 New York City	A car bomb is discovered in Times Square, New York City after smoke is seen coming from a vehicle. The bomb was ignited, but failed to detonate and was disarmed before it could cause any harm. Times Square was evacuated as a safety precaution. Faisal Shahzad pleads guilty to planting the bomb as well as 10 terrorism and weapons charges.
May 10, 2010 Jacksonville, Florida	A pipe bomb explodes while approximately 60 Muslims are praying in a mosque. The attack causes no injuries.

Date/Location	Summary of Attacks
Oct. 29, 2010	Two packages are found on separate cargo planes. Each package contains a bomb consisting of 300 to 400 grams (11–14 oz) of plastic explosives and a detonating mechanism. The bombs are discovered as a result of intelligence received from Saudi Arabia's security chief. The packages, bound from Yemen to the United States, are discovered at en route stopovers, one in England and one in Dubai in the United Arab Emirates.
Jan. 17, 2011 Spokane, Washington	A pipe bomb is discovered along the route of the Martin Luther King, Jr. memorial march. The bomb, a "viable device" set up to spray marchers with shrapnel and to cause multiple casualties, is defused without any injuries.
Sept. 11, 2012 Benghazi, Libya	Militants armed with antiaircraft weapons and rocket-propelled grenades fire upon the American consulate, killing U.S. ambassador to Libya Christopher Stevens and three other embassy officials. U.S. Secretary of State Hillary Clinton said the U.S. believed that Al Qaeda in the Islamic Maghreb, a group closely linked to Al Qaeda, orchestrated the attack.
Feb. 1, 2013 Ankara, Turkey	Ecevil Sanli detonated a bomb near a gate at the U.S. Embassy. Sanli dies after detonating the bomb. One Turkish guard is also killed. Didem Tuncay, a respected television journalist, is injured in the blast. Unlike the bombing at the embassy in Benghazi last September, the U.S. government immediately calls the bombing a terrorist attack. According to Turkish officials, the attack is from the revolutionary People's Liberation Party, which has been labeled a terrorist organization by the U.S. and other nations.
April 15, 2013 Boston, Massachusetts	Multiple bombs explode near the finish line of the Boston Marathon. Two bombs go off around 2:50 in the afternoon as runners finish the race. Three people are killed. One is an eight-year-old boy. More than 240 people are injured. Three days later, the FBI releases photos and video of two suspects in the hope that the public can help identify them. Just hours after the FBI releases the images the two suspects rob a gas station in Central Square and then shoot and kill an MIT police officer in his car. Afterward, the two men carjack an SUV and tell the driver that they had set off the explosions at the marathon. Police pursue the vehicle into Watertown. During the shootout, a MGTA officer is shot and one of the suspects identified as Tamerian Tsamaev, age 26, is killed. A suicide vest is found on his body. The other suspect Dzhokhar A. Tsamaev, age 19, remains at large for several hours, causing a massive manhunt and lockdown for all of Boson, Cambridge, and many other surrounding communities. The manhunt ends when he is found alive, but seriously injured, hiding in a boat behind a house in Watertown. The two suspects are brothers and had been living together in Cambridge. They had lived in the U.S. for about a decade, but are from an area near Chechnya, a region in Russia.

(*Continued*)

Table 2.1. (Continued)

Date/Location	*Summary of Attacks*
July 17, 2014 Ukraine	A Malaysia Airlines Boeing 777 crashes in eastern Ukraine near the Russian border, killing all 298 passengers and crew members. The crash occurs in territory where pro-Russian separatists have been battling Ukrainian troops. President Poroshenko says the crash is an act of terror. "I would like to note that we are calling this not an incident, not a catastrophe, but a terrorist act." Ukrainian and American officials say the plane is shot down by a Russian-made surface-to-air missile, citing satellite images. Poroshenko accuses the separatists of launching the missile, which they deny. Russian president Putin also denies having any role in the disaster. A day after the crash, President Obama says he believes that the rebels shot down the plane. He calls the crash a "global tragedy" and faults Putin for continuing to arm the rebels and for not stopping the lighting. Most analysts say rebels may have thought they were targeting a military transport plane rather than a commercial jet. A day before the crash, the U.S. and Europe impose further sanctions on Russia in response to Putin's refusal to stop arming the separatists.
Aug. 19, 2014	Members of ISIS behead American journalist James Foley, 40, in apparent retaliation for U.S. airstrikes against the group. Foley, who worked for Global Post, went missing in Syria in November 2012.
Sept. 2, 2014	An ISIS militant decapitates another American journalist, Steven Sotloff, 31, who worked for Time and other news outlets. He was abducted in 2013 in Syria.
Dec. 2, 2015 San Bernadino, California	Fourteen people are killed and more than 20 wounded when two people open fire at a holiday party at the Inland Regional Center, a service facility for people with disabilities and special needs in San Bernardino, California. The suspects, husband and wife Syed Rizwan Farook and Tashleen Malik, are killed in a shootout with police after the rampage. Officials say they believe the attack is terrorism related. It is the worst mass shooing in the United States since 26 people were killed in Dec. 2012 at Sandy Hook Elementary School in Newtown, Connecticut. The FBI announces that Farook and Malik had been "radicalized." The FBI also concludes that while the couple was not working with ISIS directly, their actions were inspired by the Islamic State.
June 16, 2016	A mass shooting at an Orlando nightclub in the early hours of Sunday, June 12, 2016, leaves 50 people dead, including the gunman, and more than 50 injured. The shooter is identified as Omar Mateen, 29, who entered the nightclub armed with an assault rifle and pistol. According to the authorities, Mr. Mateen pledged his allegiance to ISIS via a 911 call from inside the nightclub. This massacre is the deadliest mass shooting on U.S. soil.
March 16, 2017	Al Qaeda Operative convicted of multiple terrorism offenses targeting Americans overseas.

1. Adapted from *Terrorist Attacks* 2016. Retrieved 09/7/16 @http://www.infoplease.com/ipa/A0001454.html.

WHAT IS TERRORISM?

Since 9/11, we have heard it said by many of our teachers, neighbors, security specialists and students (and many others) that there is controversy about the definition of the politically charged word *terrorism*. Terrorism, like pollution, is a judgment call. For example, with regard to defining pollution, if two neighbors live next door to an air polluting facility, one neighbor who has no personal connection with the polluting plant is likely to label the plant's output as pollution. The other neighbor who is an employee of the plant may see the plant's pollution as dollar bills—dollars that are his or her livelihood. I have heard workers who dive into ponds full of raw sewage to find a leak in an effluent pipe say that the sewage in the pond and its associated odor is money in the bank . . . and long-term employment because toilets are not likely to vanish anytime soon. Why a money-making enterprise? Because there are few people around who would join them in diving into trenches filled with raw sewage and in doing their work. On the terrorism front, when someone deliberately spikes a tree to prevent loggers from cutting it down, the tree-spiker might feel he or she is a patriot, just a knight in shining green armor and definitely not a terrorist. On the other hand, the logger who has to take the tree down and puts his or her life and limbs at risk in taking down the spiked tree has little doubt in his or her mind on what to call the tree-spiker, and what the tree-spiker is called certainly has nothing to do with patriotism. Thus, what we are saying here in the example presented, pollution versus terrorism, is that along with attempting to define pollution, trying to define terrorism may be a judgment call, especially in the view of the terrorists.

To make the point on the countless differing views in defining terrorism, consider, for example, if we were to ask 100 different individuals to define terrorism, it is likely that we will receive a hundred different definitions. As a case in point, consider the following: if we were to ask a hundred different individuals to describe the actions of B. M. Williams VI, discussed in the prologue of this book, how would they describe her actions? You might be surprised—I was. In 2000 and again sixteen years later in 2017, pre- and post-9/11 and more recent terrorist acts, after reading B. M. Williams VI's *No Call from the Wild* and her planned actions and attitude, two groups (a 2000 and a 2017 group) of one hundred randomly selected Old Dominion University Environmental Health juniors and seniors—Generation Y and E (for entitlement) and snowflakes for those a bit different" students ranging in age from

Table 2.2. Student responses, pre- and post-9/11

	Number of Responses	
*Students' Responses Descriptors**	*Pre-9/11 (2000 Group)*	*Post-9/11 (2017 Group)*
Question 1: In your opinion, BM Williams VI was (is)		
crazy	30	0
a disgruntled former employee	10	0
insane	14	5
misguided	1	0
a cold-blooded murderer	14	16
a misfit	1	0
deranged	6	4
a lunatic	5	1
a bully	18	5
a terrorist	1	69
not sure	0	0
Totals	100	100
Question 2: In your opinion, BM Williams VI's actions are best described as		
madness	55	2
frustration	9	1
desperation	4	0
dysfunctional thinking	2	0
legitimate concern	1	0
threatening	6	1
terrorism	5	84
workplace violence	0	0
not sure	18	12
Totals	100	100

*Student response descriptors were provided to the students by the instructors.

twenty to forty years old—were asked to reply to a non-scientific survey questionnaire. The two questions and the students' responses to this unscientific survey are listed in Table 2.2.

From the Old Dominion University non-scientific survey it is clear that the students' perceptions of B. M. Williams's actions in the prologue incidents shifted dramatically from pre-9/11 to post-9/11. For example, when asked to select the best pre-9/11 descriptor to describe BM VI, "crazy" and "insane" ranked high; however, 16 years after 9/11, the students' perception shifted away from "crazy" and "insane" to "terrorist." Likewise, the students' pre-9/11 responses on BM's actions ranked high in "madness"; her actions post-9/11 overwhelmingly were described as "terrorism."

It is interesting to note that the 2000-year group reported prior to the September/October 2001 anthrax attacks and pre 9/11; the student group's responses were provided after such events as the World Trade Center attack of 1993 and Timothy McVeigh's 1995 mass murder of the occupants of the Federal Building in Oklahoma City, Oklahoma. This may help to explain why the year 2000 students were somewhat reluctant to describe B. M. VI's later actions as terrorism and/or to label her as a terrorist.

Terrorism by Any Other Name Is . . .

From the preceding discussion we might want to buy into the argument that terrorism is relative, a personal judgment. But is it really relative? And if it is relative, we must ask ourselves relative to what? Is it a personal judgment? For this text how do we define terrorism; what is terrorism? Seemingly, there is an endless list of definitions and no universally accepted definition of terrorism and other important terms. But the point is if we are going to discuss terrorism and many other aspects of critical infrastructure in a communicable manner, we need to know what the applicable terms mean.

Every branch of science, every profession, every engineering process and everything else in life has its own key concepts and language for communication. Critical infrastructure is no different. To work even at the edge of understanding critical infrastructure and how to protect it and guard against terrorism, you must acquire a fundamental knowledge of the key concepts and vocabulary related to the concepts and processes involved.

Voltaire said, "If you wish to converse with me, define your terms." In this section, we introduce the key concepts and define the terms used by critical infrastructure practitioners in applying their skills to make their technological counter-terrorism endeavors bear fruit. Although there is an extensive glossary at the end of the book, concepts and terms are presented early in the text so readers can become familiar with them now, before the text addresses the issues these concepts and terms describe. The practicing security engineer, manager, practitioner or student of protecting the country's critical infrastructure should understand these concepts and terms; otherwise, it will be difficult (if not impossible) to practice or understand the elements involved with protecting critical infrastructure.

Critical infrastructure has an extensive and unique terminology that is generally well defined, but a few terms often are not only poorly defined but also defined from different and conflicting points of view. The definition of terrorism is not an exception. The following examples make this clear.

Standard Dictionary Definition of Terrorism

After reviewing several dictionaries (too many to list here), we have found this fairly standard definition of terrorism:

> The unlawful use or threatened use of force or violence by a person or an organized group against people or property with the intention of intimidating or coercing societies or governments, often for ideological or political reasons.

America's *National Strategy for Homeland Security* defines terrorism as follows:

> Any premeditated, unlawful act dangerous to human life or public welfare that is intended to intimidate or coerce civilian populations or governments. (White House, 2006)

The U.S. State Department defines terrorism thus:

> Premeditated, politically motivated violence perpetrated against noncombatant targets by subnational groups or clandestine agents. (U.S. Congress, 2005)

The FBI definition of terrorism is as follows:

> The unlawful use of force or violence against persons or property to intimidate or coerce a Government, the civilian population, or any segment thereof, in furtherance of political or social objectives. (FBI, 2006)

Note that the FBI divides terrorism into two categories: domestic (homegrown), involving groups operating in and targeting the U.S. without foreign direction; and international, involving groups that operate across international borders and/or have foreign connections.

Well, at this point the obvious question is: Do you now know what terrorism is? That is, can you definitively define it? If you can't define it, you are not alone—not even the U.S. government can definitively define it. Maybe we need to look at other sources—views from the real experts on terrorism.

Osama bin Laden's View on Terrorism

> Wherever we look, we find the U.S. as the leader of terrorism and crime in the world. The U.S. does not consider it a terrorist act to throw atomic bombs at nations thousands of miles away [Japan during World War II], when those bombs would hit more than just military targets. Those bombs rather were thrown at entire nations, including women, children, and elderly people. (Bergen, 2002, 21–22)

Another View

This view is from court testimony on terrorism from Ramzi Ahmed Yousef, who helped organize the first terrorist attack on the World Trade Center:

> You keep talking also about collective punishment and killing innocent people to force governments to change their policies; you call this terrorism when someone would kill innocent people or civilians in order to force the government to change its policies. Well, when you were the first one who invented this. . . .
>
> You were the first one who killed innocent people, and you are the first one who introduced this type of terrorism to the history of mankind when you dropped an atomic bomb which killed tens of thousands of women and children in Japan and when you killed over a hundred thousand people, most of them civilians, in Tokyo with fire bombings.
>
> You killed them by burning them to death. And you killed civilians in Vietnam with chemicals as with the so-called Orange agent. You killed civilians and innocent people, not soldiers, innocent people every single war you went. You went to wars more than any other country in this century, and then you have the nerve to talk about killing innocent people.
>
> And now you have invented new ways to kill innocent people. You have so-called economic embargo which kills nobody other than children and elderly people, and which other than Iraq you have been placing the economic embargo on Cuba and other countries for over 35 years. . . .
>
> The government in its summations and opening said that I was a terrorist. Yes, I am a terrorist and I am proud of it. And I support terrorism so long as it was against the United States Government and Israel, because you are more than terrorists; you are the one who invented terrorism and using it every day. You are butchers, liars and hypocrites. (Excerpts from court 1998)

And finally, here is an old cliché on a terrorist:

> One man's terrorist is another man's freedom fighter.

Again, from the preceding points of view, it can be seen that defining terrorism or the terrorist is not straightforward and never easy. Even the standard dictionary definition leaves us with the vagaries and ambiguities of other words typically associated with terrorism, such as in the definitions of *unlawful* and *public welfare* (Sauter and Carafano, 2005).

Raphael Perl (2004) in a Congressional Research Service (CRS) report points out that one definition widely used in U.S. government circles, and incorporated into law, defines *international terrorism* as terrorism involving the citizens or property of more than one country.

Terrorism is broadly defined as politically motivated violence perpetrated against noncombatant targets by subnational groups or clandestine agents. For example, kidnapping of U.S. birdwatchers or bombing of U.S.-owned oil pipelines by leftist guerrillas in Columbia would qualify as international terrorism. In 22 U.S.C. 2656f, a *terrorist group* is defined as a group which practices terrorism or has significant subgroups that practice terrorism. Perl (2004) points out that one of the shortfalls of this traditional definition is its focus on groups and its exclusion of individual ("lone wolf") terrorist activity, which has recently risen in frequency and visibility.

At this point, readers may wonder, "Why should we care?—that is, what difference does it make (do you remember this infamous statement?)—as to what the definition of terrorist or terrorism is?" Definitions are important because in order to prepare for the terrorism contingency, domestic or international, we must have some feel, as with any other problem, for what it is we are dealing with. We are fighting a war of ideas. We must attempt to understand both sides of the argument, even though the terrorist's side makes no sense to an American or other freedom-loving occupant of the globe.

The *Washington Times* (2009) reports that in an interview with German magazine *Der Spiegel*, the Secretary of Homeland Security said that she's rejected the word "terrorism," and has instead renamed terrorist acts using the euphemism "man-caused disasters," because" it demonstrates that we want to move away from the politics of fear, toward a policy of being prepared for all risks that can occur. This is analogous to stating that we should refer to serial killers, serial rapists, and serial arsonists as human-caused afflictions. After all, we do not want to create fear about serial offenders. Get Real! The author of this text suggests that former Secretary Napolitano read the remarks we mentioned earlier by Ramzi Ahmed Yousef:

> Yes, I am a terrorist and I am proud of it. And I support terrorism as long as it was against the United States Government and Israel.

Finally, while it is difficult to pinpoint an exact definition of terrorism (but not difficult to rename it and call it something else), we certainly have little difficulty in identifying it when we see it, when we feel it, when we suffer from it. By any other name terrorism is best summed up as an absolute feeling of Terror—nothing judgmental about that—just Terror with a capital *T*.

Terror!

Ok, you say, several definitions of terror have been presented. The question is which definition will be used in this book and in all 16 volumes of the Critical Infrastructure series? Probably the best definition of terrorism is given in United States Department of Homeland Security's *Defense Industrial Base Sector-Specific Plan* (2010). DHS's definition:

> Terrorism is the calculated use of unlawful violence or threat of unlawful violence to inculcate [instill] fear, intended to coerce or intimidate governments or societies in the pursuit of goals that are generally political, religious, or ideological.

In addition to the definition of terrorism a few of the important definitions related to this discussion of critical infrastructure are included in the following (DHS, 2010).

All hazards—A grouping classification encompassing all conditions, environmental or manmade, that have the potential to cause injury, illness, or death; damage to or loss of equipment, infrastructure services, or property; or alternatively causing functional degradation to social, economic, or environmental aspects.

Asset (Infrastructure)—A distinguishable network entity that provides a service or capability. Assets are people, physical entities, or information located either within or outside the United States and owned or operated by domestic, foreign, public, or private sector organizations. DHS states an asset is a person, structure, facility, information, material, or process that has value.

Business continuity—Strategic and tactical capability of the organization to plan for and respond to incidents and business disruptions and continue business operations at an acceptable predefined level. DHS states business community is the ability of an organization to continue to function before during, and after a disaster.

Consequence—The effect of an event, incident, or occurrence. For the purposes of this book, consequences are divided into four main consequences: public health and safety, economic, psychological, and governance impacts.

Control systems—Computer-based systems used within many infrastructures and industries to monitor and control sensitive processes and physical functions. These systems typically collect measurements and operational data from the field, process and display the information, and relay control commands to local or remote equipment or human-machine interfaces (operators). Examples of types of control systems include Supervisory Control and Data Acquisition Systems, Process Control Systems, and Distributed Control Systems.

Critical infrastructure—Systems and assets, whether physical or virtual, so vital that the incapacity of destruction of them may have a debilitating impact on the security, economy, public health or safety, environment, or any combination of these factors, across any federal, state, regional, territorial, or local jurisdiction.

Critical infrastructure information (CII)—Information that is not customarily in the public domain and is related to the security of critical infrastructure or protected systems, CII consists of records and information concerning any of the following:

- Actual, potential, or threatened interference with, attack on, compromise of, or incapacitation of critical infrastructure or protected systems by either physical or computer-based attack or other similar conduct (including the misuse of or unauthorized access to all types of communications and data transmission systems) that violates federal, state, or local law; harms the interstate commerce of the United States; or threatens public health or safety.

- The ability of any critical infrastructure or protected system to resist such interference, compromise, or incapacitation, including any planned or past assessment, projection, or estimate of the vulnerability of critical infrastructure of a protected system, including security testing, risk evaluation thereto, risk management planning, or risk audit.
- Any planned or past operation problem or solution regarding critical infrastructure or protected systems, including repair, recovery, insurance, or continuity, to the extent that it is related to such interference, compromise, or incapacitation.

Cybersecurity—The prevention of damage to, unauthorized use of, or exploitation of, and, if needed, the restoration of electronic information and communication systems and the information contained therein to ensure confidentiality, integrity, and availability. Includes protection and restoration, when needed, of information networks and wire line, wireless, satellite, public safety answering points, and 911 communication systems and control systems.

Defense critical asset—An asset of such extraordinary importance to DoD operations in peace, crisis, and war that its incapacitation or destruction would have a very serious, debilitating effect on the ability of DoD to fulfill its mission.

Defense Production Act of 1950 (DPA)—Under Title 1 of the act, as amended (50 U.S.C. App. 2061 et seq.), the president is authorized to require preferential acceptance and performance of contracts or orders supporting certain approved national defense and energy programs, and to allocate materials, services, and facilities in such a manner as to promote these approved programs. Under Title VII, the president is authorized to conduct mandatory surveys and analyses, and prepare reports on specific subsectors of the U.S. DIB. The DPA's current definition of "national defense" includes programs for military and energy production or constructions, military or critical infrastructure assistance to any foreign nation, homeland security, stockpiling, space, and any directly related activity. The DPA authority has also been extended to support emergency preparedness activities under Title VI of the

Robert T. Stafford Disaster Relief and Emergency Assistance Act (the Stafford Act), as amended (42 U.S.C 5195 et seq.), and critical infrastructure protection and restoration (Public Law 111-67).

Dependency—The one-directional reliance of an asset, system, network, or collection thereof, within or across sectors, on input, interaction, or other requirement from other sources in order to function properly. A relationship or connection whereby one entity is influenced or controlled by another (Department of Defense Directive (DoDD) 3020.40, 19 Aug 2005).

- **Interdependency**. Relationships or connections between entities of different functions, networks, sectors, or service (DoDD 30020.40, 19 Aug. 2005).
- **Intra-dependency**. Relationships or connections between entities within a common function, network, sector, or service (DoDD 3020.40, 19 Aug. 2005).

DIB Critical Asset—An asset determined by DoD to be critical to successful execution of its missions.

Government Coordinating Council (GCC)—The government counterpart to the SCC for each sector established to enable interagency coordination. The GCC is comprised of representatives across various levels of government (federal, state, local, tribal, and territorial) as appropriate to the security and operational landscape of each individual sector.

Incident—An occurrence caused by either human action or natural phenomena that require action to prevent or minimize loss of life or damage to property and/or natural resources (JP 3-18).

Infrastructure—(1) The framework of networked assets that comprise identifiable industries, institutions, or distribution capabilities that enable continued flow of goods and services. (DoDD 3020.40). All building and permanent installations necessary for support, redeployment, and military forces operations (e.g., barracks, headquarters, airfields, communications facilities, stores, port installations, and maintenance stations) (JP 3-35). (2) The framework of interdependent networks and systems comprising identifiable industries, institutions (including people and procedures), and distribution capabilities that provide a reliable flow of products and services essential to the defense and economic security of the United States, the smooth functioning of government at all levels, and society as a whole.

Consistent with the definition in the Homeland Security Act of 2002, infrastructure includes physical, cyber, and/or human elements.

Key resources—Publicly or privately controlled resources essential to the minimal operation of the economy and government.

Mitigation—Actions taken in response to a warning or after an incident occurs that are intended to lessen the potentially adverse effects on a given military operation or infrastructure (DoDD 3020.40).

Network—A group or system of interconnected or cooperating entities, normally characteristic as being nodes (assets), and the connections that link them (DoDD 3020.40). The DHS definition of network is a group of components that share information or interact with each other in order to perform a function.

Owners and operators—Those entities responsible for day-to-day operation and investment in a particular asset or system.

Preparedness—Activities necessary to build, sustain, and improve readiness capabilities to prevent, protect against, respond to, and recover from natural or manmade incidents. Preparedness is a continuous process involving efforts at all levels of government, and between government and the private sector and nongovernmental organizations to identify threats, determine vulnerabilities, and identify required resources to prevent, respond to, and recover from major incidents.

Prevention—The security procedures undertaken by the public and private sectors to discourage terrorist acts (JP 3-07.2).

Prioritization—In the context of this presentation, prioritization is the process of using risk assessment results to identify where risk-reduction or mitigation efforts are most needed, and subsequently determine which protective actions should be instituted in order to have the greatest effect.

Protection—Preservation of the effectiveness and survivability of mission-related military and nonmilitary personnel, equipment, facilities, information, and infrastructure deployed or located within or outside the boundaries of a given operational area (JP 3-0). DHS states that protection is an action or measure taken to cover or shield from exposure, injury, or destruction. In the context of this book, protection includes actions to deter the threat, mitigate the vulnerabilities, or minimize the consequences associated with a terrorist attack or other incident. Protection can include a wide range of activities, such as hardening facilities, building resilience and redundancy, incorporating hazard resistance into facility design, initiating active or passive countermeasures, installing security systems, promoting workforce surety, training and exercise, and implementing cybersecurity measures, among various others.

Recovery time objective—Target time set for resumption of product, service, or activity delivery after an incident. Note: The recovery time objective must be less than the maximum tolerable period of disruption.

Resilience/resiliency—The adaptive capacity of an organization in a complex and changing environment (ISO 31000).

Note 1: Resilience is the ability of an organization to resist being affected by an event or the ability to return to an acceptable level of performance in an acceptable period after being affected.

Note 2: Resilience is the capability of a system to maintain its functions and structure in the face of internal and external change, and to degrade gracefully when it must.

Note 3: There are three levels of resilience: National, Continuous Integration (CI System), and Individual.

- National—Willpower to bounce back
- CI System—Identify critical nodes for protection and planning prioritization; create adaptive grids to ensure redundancy
- Individual

Response—Activities that address the short-term, direct effects of an incident, including immediate actions to save lives, protect property, and meet basic human needs. Response also includes the execution of emergency operations plans and incident mitigation activities designed to limit the loss of life, personal injury, property damage, and other unfavorable outcomes. As indicated by the situation, response activities include applying intelligence and other information to lessen the effects or consequences of an incident; increasing security operations; continuing investigations into the nature and source of the threat; ongoing surveillance and testing processes; immunizations, isolation, or quarantine; and specific law enforcement operations aimed at preempting, interdicting, or disrupting illegal activity and apprehending actual perpetrators and bringing them to justice.

Risk—Probability and severity of loss linked to threats or hazards. (DoDD 3020.40) DHS states risk is the potential for an unwanted outcome resulting from an incident, event, or occurrence, as determined by is likelihood and the associated consequences.

Risk management framework—A planning methodology that outlines the process for setting goals and objectives; identifying assets, systems, and networks; assessing risks; prioritizing and implementing protection programs and resilience strategies; measuring performance; and taking corrective action. Public and private sector entities often include risk management frameworks in the business continuity plans.

Sector—A sector is a logical collection of assets, systems, or networks that provide a common function to the economy, government, or society.

System—A functionally, physically, and/or behaviorally related group of regular interaction or interdependent elements, that group of elements forming a unified whole (JP 3-0). DHS states a system is a any combination of facilities, equipment, personnel, procedures, and communications integrated for a specific purpose.

Threat—An adversary having the intent, capability, and opportunity to cause loss or damage (DoDD 3020.40). DHS definition of threat is a natural or manmade occurrence, individual, entity, or action that has or indicates the potential to harm life, information, operations, the environment, and/or property.

Value proposition—A statement that outlines the national and homeland security interest in protecting the nation's critical infrastructure and articulates benefits gained by all partners through the risk management framework.

Vulnerability—The characteristic of an installation, system, asset, application, or its dependencies that could cause it to suffer a degradation or incapacity to perform its designated function as a result of having been subjected to a certain level of threat or hazard (DoDD 3020.40). The DHS definition of vulnerability is a physical feature or operational attribute that renders an entry open to exploitation or susceptible to a given hazard.

REFERENCES AND RECOMMENDED READING

Bergen, P. L. (2002). *Holy War, Inc: Inside the Secret World of Osama bin Laden.* New York: Touchstone Press, 21–22.

CRS. (2006). *Chemical Facility Security. CRS Report for Congress.* Washington, DC: Congressional Research Service—The Library of Congress.

DHS. (2003). The National Strategy for the Physical Protection of Critical Infrastructures and Key Assets. https://www.dhs.gov/xlibrary/assets/physical_Strat.

DHS. (2010). Defense Industrial Base Sector-Specific Plan. Washington, DC: U.S. Department of Homeland Security.

DHS. (2017). *Communications Sector*. Washington, DC: Department of Homeland Security. Accessed 04/06/2017. https://www.dhs.gov/commuincaitns-sector.

FEMA. (2015). *Protecting Critical Infrastructure Against Insider Threats*. Accessed 17 April 2015. http://emilms.fema.gov/IS0915/IABsummary.htm.

Gurr, T. R. (1968). Psychological Factors in Civil Violence. *World Politics*, 20, no. 32 (January): 245–78.

Haimes, Y. Y. (2004). *Risk Modeling, Assessment, and Management*, 2nd Edition. New York: John Wiley & Sons, 699.

Henry, K., (2002). New face of security. *Gov. Security*, Apr., pp. 30–37.

Lindsey, H., (2001). *Vocabulary of Hate*. Accessed 04/18/08. www.wordnetdaily.com.

New York Times. (1998). Excerpt from court testimony. Page B4.

NSHS. (2006). *National Strategy for Homeland Security*. Accessed 5/13/06. www.whitehouse/homeland.

Old Dominion University. (2000; 2002). *Violence in the Workplace: Security Concerns*. From a series of lectures presented to environmental health students. Norfolk, Va.

Perl, R. (2004). *Terrorism and National Security: Issues and Trends*. CRS Issue Brief IB10119. Washington, DC.

Sauter, M. A., and Carafano, J. J. (2005). *Homeland Security: A Complete Guide to Understanding, Preventing, and Surviving Terrorism*. New York: McGraw-Hill.

Spellman, F. R. (1997). *A Guide to Compliance for Process Safety Management/Risk Management Planning (PSM/RMP)*. Lancaster, PA: Technomic Publishing Company.

The Washington Times. (2009). Napolitano tells it like it isn't. Retrieved March 29, 2009. www.washingtontimes.com/news/2009/mar/29/tell-it-like-it-is-man-casued-disasters-is-mapolit/.

United States Congress. (2005). Annual Country reports on terrorism. 22 USC, chapter 38, section 2656f.

3

Critical Infrastructure

In providing a secure environment, the key function is to connect the dots. In designing and building infrastructure to withstand terrorism, the key word is Resilience, Resilience, Resilience.

—Frank R. Spellman

While it is not so easy to definitively define terrorism and/or the terrorist, we have less difficulty identifying the likely targets of terrorists. In America, we call these likely targets our critical infrastructure.

So in war, the way to avoid what is strong and to strike at what is weak.

—Sun Tzu

WHAT IS CRITICAL INFRASTRUCTURE?

For the United States of America, 9/11 was a slap in the face, a punch in the gut (actually, the ultimate sucker punch), and a most serious wake-up call. The 9/11 wake-up call generated several reactions on our part—obviously, protecting ourselves from further attack became (and hopefully still is) priority number one. In light of this important need (i.e., the survival of our way of life), the Department of Homeland Security was created. According to Governor Tom Ridge, "You may say Homeland Security is a Y2K problem that doesn't end Jan. 1 of any given year" (Henry, 2002). And, according to

Barack Obama (2007), the Department of Homeland Security does "the work that ensures no other family members have to lose a loved one to a terrorist who turns a plane into a missile, a terrorist who straps a bomb around her waist and climbs aboard a bus, a terrorist who figures out how to set off a dirty bomb in one of our cities."

Among other things, the new emphasis on homeland security pointed to the need to protect and enhance the security of the nation's critical infrastructure. Critical infrastructure can be defined or listed in many ways. Generally, governments use the term to describe material assets, systems, and services that are essential for the functioning of an economy and society and maintaining public confidence. Destruction or compromise of any of these systems or services would have a debilitating impact on the area either directly, through interdependencies or from cascading effects. For the purpose of this text, critical infrastructure is defined as those assets of physical, key resources, and computer-/service-based systems that are essential to the minimum operations of the economy and government. The sixteen Critical Infrastructures (used to be fourteen but was updated in February 2013 and now consists of sixteen) making up volumes in this series are the following:

- Critical manufacturing
- Food and agriculture
- Financial (banking and finance)
- Chemical and hazardous materials
- Defense industrial base
- Emergency services
- Energy
- Pubic health
- Communications
- Transportation
- Water and wastewater
- Dams
- Information technology
- Commercial facilities
- Government Facilities
- Nuclear power plants

DID YOU KNOW?

Over 85 percent of the critical infrastructure within the U.S. is owned and operated by the private sector.

Although we did not list cyberspace and all ancillaries (excluding, of course, the listing of information technology) involved in or with digital operations (e-technology), in this current era we can state without equivocation, doubt, ambiguity, or vagueness that the digital communication connection is the glue that holds all critical infrastructure together. This is the case, of course, because all separate infrastructures are interconnected (lots of crossover; refer to Figure 3.1) in one way or another. This may surprise you to some degree, but think about it—we are not speaking about rocket science here; no, instead, we are speaking about the present reality of e-technology. It would be hard to imagine that any of the above listed infrastructure sectors could operate today without e-technology, without some sort of communication.

For example, even though many of us did it in the past, today how would we go about withdrawing money from the bank without e-banking technology? Today we can conduct our banking at any time we wish, from just about any location in the world. Modern life without debit cards and ATMs would be a rude awakening for many of us.

We could go down the list of critical infrastructures and easily point out where e-technology and the specific industry interface. However, since this is a discussion focusing on the communications sector, we will keep the focus all inclusive on e-technology as it relates or interfaces with the communications sector. If people are not familiar with the communications sector beyond the previous brief introduction, description, and discussion, Table 3.1 presents a list of the components that make up the communications sector, and subsequent detailed discussion of each sector component follows.

Table 3.1. Communications Sector Components

Components
Wireline Infrastructure
Wireless Infrastructure
Satellite Infrastructure
Cable Infrastructure
Broadcasting Infrastructure

WIRELINE INFRASTRUCTURE

Although we are experiencing an ongoing revolution in communications technology development (i.e., the communications sector transition to packet-based networks), Public Switched Telephone Network (PSTN), as well as enterprise networks, the traditional wireline infrastructure remains the backbone of the communications infrastructure. The PSTN is a domestic communications network accessed by telephones, key telephone systems, private branch exchange (PBX) trunks, and data arrangements. The completion of the circuit between the call originator and the call receiver requires network signaling in the form of dial pulses or multi-frequency tones. These components are connected by nearly 2 billion miles of fiber and copper cable (physical), have IT systems that monitor and move the data (cyber), and have dedicated staff to ensure service (people).

DID YOU KNOW?

An enterprise network is an enterprise's communications backbone that helps connect computers and related devices across departments and workgroup networks, facilitating insight and data accessibility. An enterprise network reduces communication protocols, facilitating system and device interoperability, as well as improved internal and external enterprise data management (Techopedia, 2017).

The wireline component has traditionally been divided between Interchange Carriers (IXCs) and Local Exchange Carriers (LECs). Local access and transportation areas (LATAs) provide definition to the areas of provisioning responsibilities. Generally, the incumbent LEC companies provided local and intra-LATA toll services, with the IXCs providing interexchange toll services, with the ISCs providing interexchange tool services. However, regulatory developments have blurred the lines between those providers. Now, many traditional LECs are evolving to provide long-distance services, and more ISCs are becoming full-service providers. In addition, following passage of the Telecommunications Act of 1996, new Competitive Local Exchange Carriers (CLECs) entered the local long-distance and data services markets, as did some tradition Cable Television (CATV) providers. Through their wireline networks, IXCs, LECs, and CLECs are also leading providers of Internet access and broadband services. Future providers may involve non-traditional platforms and infrastructure, such as broadband over power lines (DHS, 2010).

Key wireline network and transmission elements include the following, many of which are detailed in Figure 3.1:

- *Local Exchange Switching*: The traditional local exchange network is a hierarchical structure with end-to-end connections using customer, local, and long-distance networks. This, coupled with an ability to concentrate more traffic over few links, has lowered the cost of long-distance traffic to the consumer. However, these same capabilities have resulted in more significant impacts when long-distance links and node and link failures occur.
- *Interexchange Switching*: The traditional interexchange networks are independent mesh structures that incorporate direct point-to-point connections between nodes.
- *Transmission Links*: The physical unit of a subnetwork that provides the transmission connection between adjacent nodes.

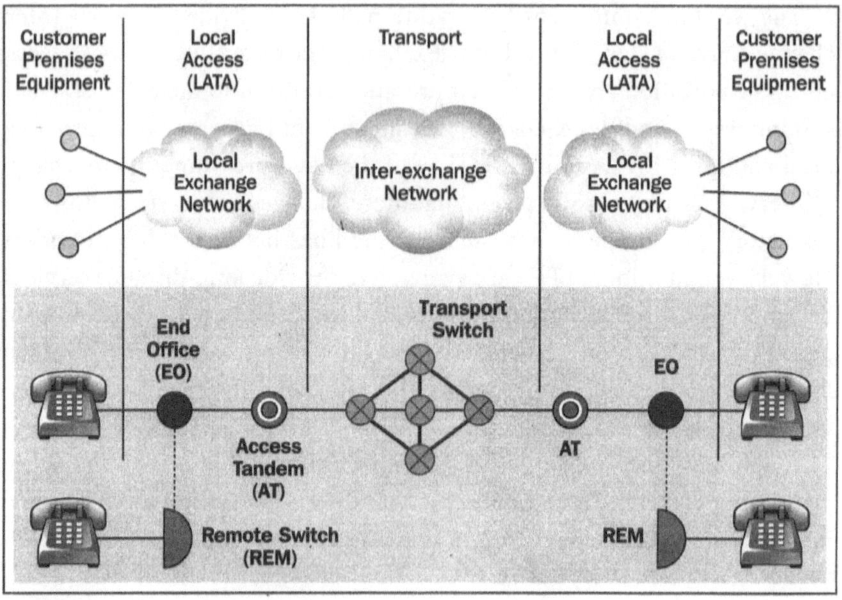

FIGURE 3.1
Wireline Network Architecture.

Source: DHS (2010). Communications Sector-Specific Plan. Washington, DC: U.S. Department of Homeland Security.

Signaling System Number 7 (SS7)

SS7 is a communications protocol developed in 1975 that provides signaling and control for various network services and capabilities. SS7 networks are medium-speed (56 or 64 kilobits per second), packet-switched networks that overlay the carriers' circuit-switched networks and provide network control functionality to the PSTN. SS7 is composed of a series of interconnected network elements (e.g., switches, databases, and routing nodes). The SS7 protocol also has significant cyber implications because it affords the interface from circuit-switched (traditional) networks to IP-based networks. Figure 3.2 illustrates how the SS7 connects to the wireline network (DHS, 2010).

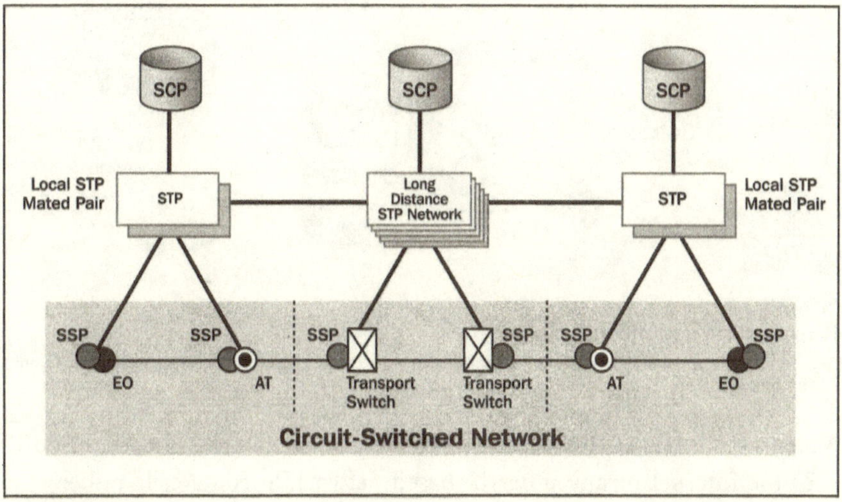

FIGURE 3.2
SS7 and Wireline Network Architecture.
Source: DHS (2010). Communications Sector-Specific Plan. Washington, DC: U.S. Department of Homeland Security.

Next-Generation Networks (NGNs)

The concept of NGNs considers new realities within the communications industry and can be defined as high-speed, converged circuit-switched and packed-switched and networks capable of routing a multitude of services, including voice, data, video, and multimedia, across variant platforms. NGNs leverage open architecture over a common transport network with an emphasis on optical networking and intelligent or NGN "aware" elements. NGNs seamlessly blend the PSTN and the packet-switched data network and are also called converted networks because they integrate voice and data communications across traditionally divergent fixed and mobile platforms to an increasing array of end-use devices.

Key NGN functional elements (see Figure 3.3) include the media gateway (MG), the signaling gateway (SG), and the media gateway controller (MGC).

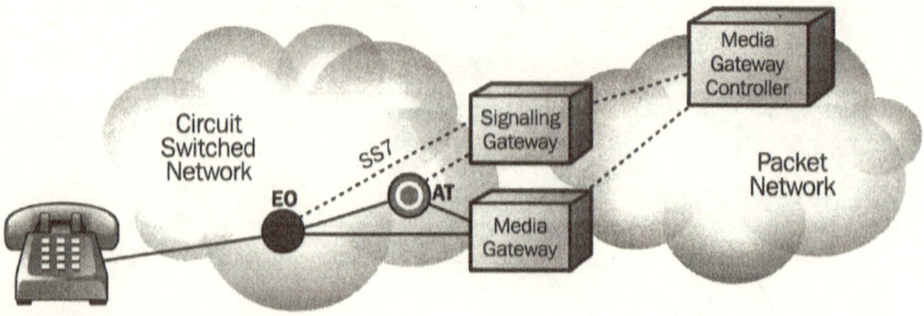

FIGURE 3.3
NGN functional elements.

Source: DHS (2010). Communications Sector-Specific Plan. Washington, DC: U.S. Department of Homeland Security.

Voice over Internet Protocol (VoIP). Voice over Internet Protocol (VoIP) uses the Internet or any other IP-based network to route calls rather than the PSTN with a packet-switched network being used rather than dedicated, circuit-switched telephone transmission lines. Figure 3.4 depicts the elements of a VoIP network configuration.

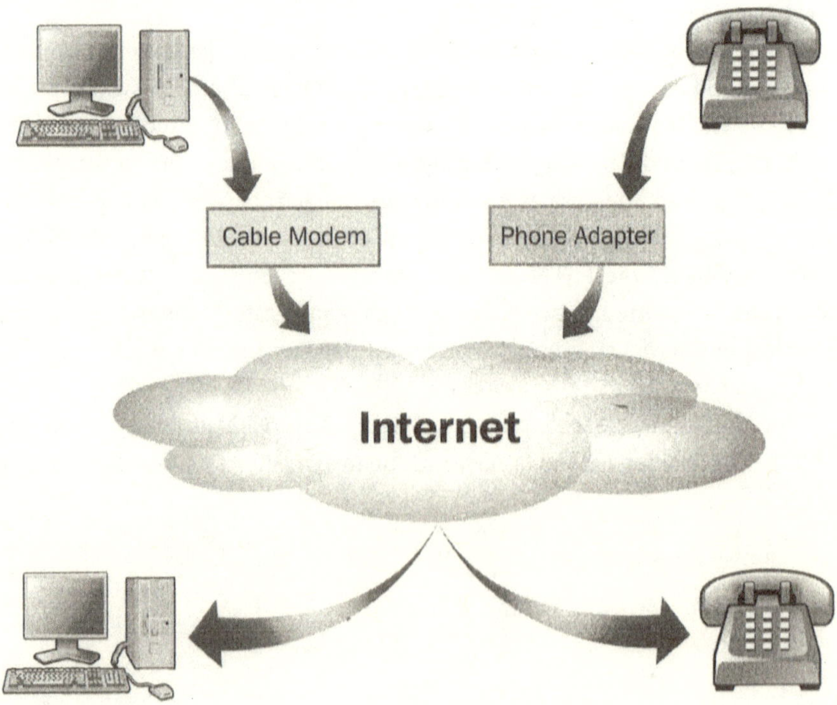

FIGURE 3.4
VoIP Network.

Submarine Cable Networks. Submarine cable networks are long-haul wireline networks constructed in major bodies of water to interconnect networks. Key submarine cable components (see Figure 3.5) include landing locations/cable heads and switching centers.

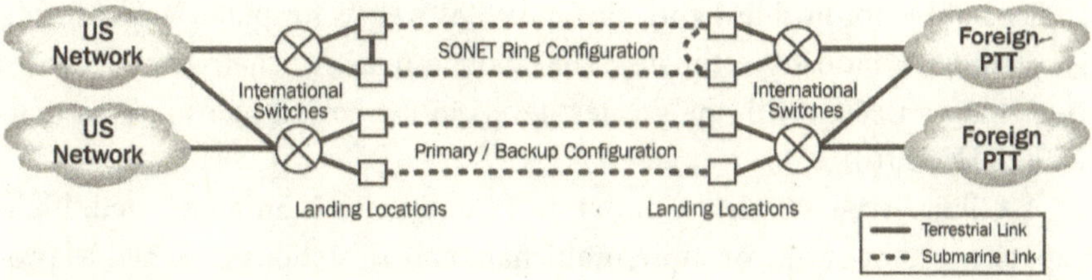

FIGURE 3.5
Submarine cable architecture.

Source: DHS (2010). Communications Sector-Specific Plan. Washington, DC: U.S. Department of Homeland Security.

The Internet. The Internet includes the global infrastructure of packet-based networks and databases that use a common set of protocols for communication (see Figure 3.6). The networks are connected by various transports. The most common examples of Internet access include ordinary telephone lines (dialup); broadband services such as Digital Subscriber Lines (DSL)and cable modems; Integrated Services Digital Network (ISDN); T1 and T3 lines; and interconnected wireless services, infrastructure, and devices (DHS, 2010).

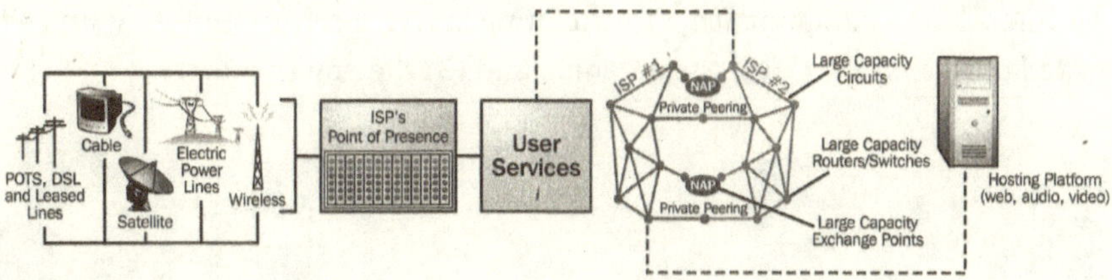

FIGURE 3.6
Internet architecture.

Source: DHS (2010). Communications Sector-Specific Plan. Washington, DC: U.S. Department of Homeland Security.

WIRELESS INFRASTRUCTURE

A wireless network is a computer network that uses wireless data connections between network nodes and includes cellular phone, paging, personal communications services, high-frequency radio, unlicensed wireless, and other commercial and private radio services. Wireless services have become indispensable for business and consumers, as well as for public safety needs. According to industry estimates, U.S. mobile market penetration exceeded two-thirds in 2005, with the greatest levels in the largest metropolitan markets (DHS, 2010).

A cellular-type Wireless Communication System is an automated, high-capacity system of one or more multichannel base stations designed to provide radio communications services to users over a wide area in a spectrally efficient manner. A cellular-type architectural system operates by dividing a large geographical service area into cells and assigning the same channels to multiple, nonadjacent cells. This design allows channels to be reused, increasing spectrum efficiency. As a subscriber travels across the service area, the call is transferred (handed off) from one cell to another without noticeable interruption. Cellular-type wireless networks are composed of several elements (see Figure 3.7) including cell sites, mobile switching centers (MSC) and base switching controllers (BSC).

High-frequency (HF) radio (commonly known as short-wave radio) can be used for communication over great distances and between points separated by geographic barriers (e.g., mountains). An HF radio system consists of three basic components: (1) the transmitter/receiver unit (commonly called the transceiver), (2) the antenna, and (3) the power source.

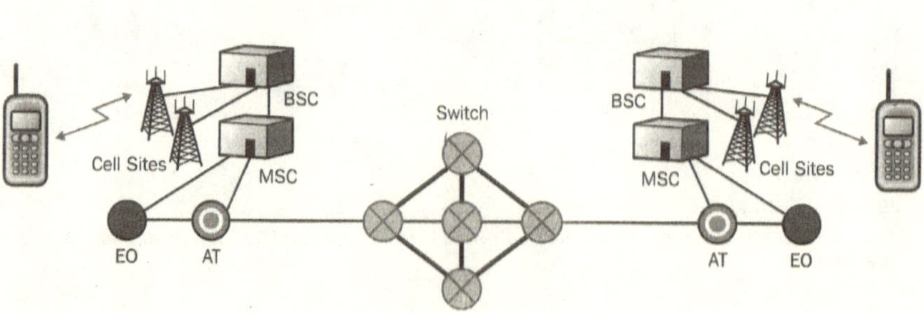

FIGURE 3.7
Wireless Network Architecture.

Source: DHS (2010). Communications Sector-Specific Plan. Washington, DC: U.S. Department of Homeland Security.

SATELLITE INFRASTRUCTURE

Satellite communications systems deliver data, voice, and video services. Networks may be private and independent of the terrestrial infrastructure or may share common facilities (e.g., a Teleport) and be combined with terrestrial services to deliver information to the intended recipient(s).

Geostationary Earth Orbit (GEO) systems typically required a single satellite for a continental United States (CONUS) footprint and three satellites to have a global footprint. Non-geostationary Low Earth Orbit (LEO) requires numerous satellites for global coverage. A group of artificial satellites working in concert is thus known as a satellite constellation with coordinated ground coverage, synchronized so that they overlap well in coverage (see Figure 3.8). Satellite services exist in several forms:

- Fixed
- Transportable
- On the move
- Handheld (Mobile Satellite Services (MSS))
- Broadcast Satellite Services (BSS (public broadcast)).

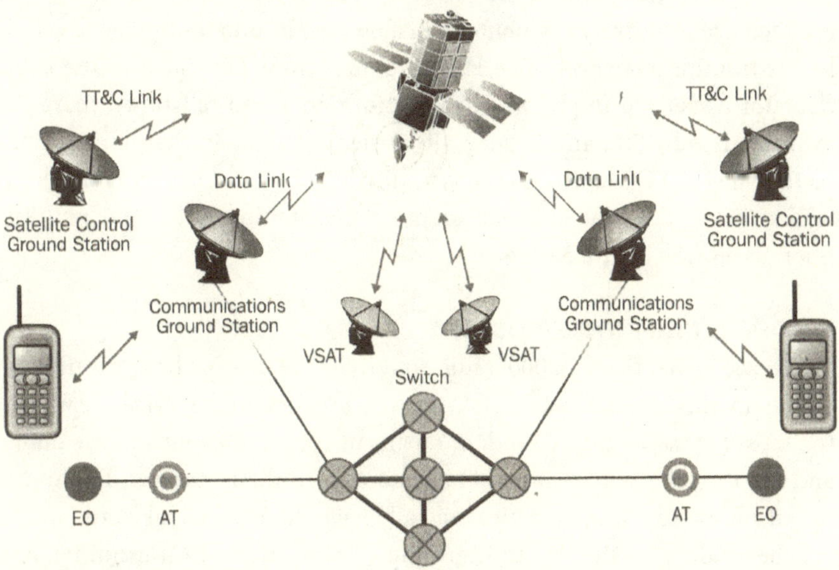

FIGURE 3.8
Satellite Network Architecture.

Source: DHS (2010). Communications Sector-Specific Plan. Washington, DC: U.S. Department of Homeland Security.

The technology can deliver two-way point-to-point or mesh converged communications (voice, data, and video) for access and/or backbone services, as well as multicast (video and or data) for distribution of content to large geographically dispersed audiences. MSS services are primarily vice with limited to mid-range data capabilities.

Important satellite network components include ground stations; telemetry, tracking, and command links (TT&Cs); very small aperture terminals (VSATs), and data links (see Figure 3.8).

CABLE INFRASTRUCTURE

The United States CATV industry serves approximately 63.1 million basic cable customers, representing a market penetration of 50.3 percent of U.S. television households. The cable industry is composed of approximately 7,791 cable systems (DHS, 2010). Many of the systems are owned by Multiple-Systems Operators (MSOs), providing services to about 58 million households. Cable services include analog and digital video programming services, and high-speed Internet access service.

Most cable systems use a mixture of fiber and coaxial cable that provides bidirectional signal paths to the customer. This Hybrid Fiber Coaxial (HFC) architecture effectively segments the cable system into a number of parallel distribution networks. The HFC architecture is beneficial to the cable operator because it improves signal performance and reliability, increases available bandwidth, and is generally easier to maintain than older architecture. The HFC architecture is typically based on a three-level topology, which includes a headend, one or more distribution hubs, and multiple fiber nodes (see Figure 3.9).

BROADCASTING INFRASTRUCTURE

There are more than 14,000 radio and 1,700 television broadcasting facilities in the United States. As a group, broadcasting services represent the most pervasive and redundant means of communications to the public and are the only services available at no cost to the consumer. Radio and television broadcasting facilities typically have three principal components: (1) the studio, (2) the transmitter, and (3) the studio-to-transmitter link (STL). Broadcasters typically design multiple levels of redundancy into each of these components, including:

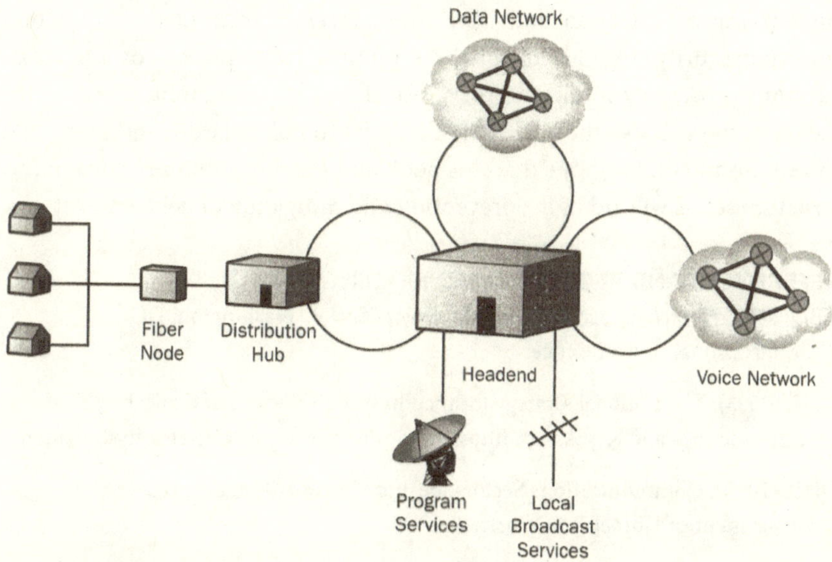

FIGURE 3.9
Cable Network Architecture.

Source: DHS (2010). Communications Sector-Specific Plan. Washington, DC: U.S. Department of Homeland Security.

- Main and back-up studio facilities
- Multiple, independent STLs utilize wireline, wireless, and satellite radio channels
- Primary and auxiliary transmitters, towers, and antennas.

Provisions are typically made for backup electrical power using battery backup and diesel/gasoline-powered generators.

THE BOTTOM LINE

Again, it is important to point out that the Communications Sector provides services that are essential to the U.S. defense, offense, security, economy and to protect and ensure the so-called "good life" (standard of living) we presently enjoy. The Communications Sector is well aware of its vulnerabilities and is leading a significant mandated and voluntary effort to increase its planning and preparedness. Cooperation through industry groups has resulted in substantial information sharing of effective and best practices across the sector. Many sector owners and operators have extensive experience with infrastructure protection and have more recently focused their attention on cybersecurity. In addition to the economic consequences of a successful home-grown or foreign terrorist attack against communications sector facilities, there is also the potential of a threat to public health and safety and the environment. We hope that this book and the others in the critical infrastructure series will aid in the prevention and mitigation of deliberate attacks.

REFERENCES AND RECOMMENDED READING

CBO. 2004. *Homeland Security and the Private Sector*. Washington, DC: Congressional Budget Office.

DHS. (2003). The National Strategy for the Physical Protection of Critical Infrastructures and Key Assets. https://www.dhs.gov/xlibrary/assets/physical_Strat.

DHS. (2010). Communications Sector-Specific Plan. Washington, DC: U.S. Department of Homeland Security.

FAO. (2005). *Bridging the Rural Digital Divide*. New York: United Nations. Accessed 04/19/08. www.fao.org/rdd.

FEMA. (2015). *Protecting Critical Infrastructure Against Insider Threats*. Accessed 17 April 2015. http://emilms.fema.gov/IS0915/IABsummary.htm.

FR. (2003). Notice of Proposed Rulemaking. *Federal Register*, vol. 68, no. 90.

Henry, K. (2002). New face of security. *Gov. Security*, Apr., pp. 30–37.

Horn, F. P. (1999). Statement Made before the United States Senate Emerging Threats and Capabilities Subcommittee of the Armed Services Committee. Accessed 06/27/07. www.Senate.gov/~armed_servies/statement/1999/991027fh.pdf.

Lane, J. (2002). Sworn Testimony, Congressional Field Hearing, House Committee on Government Reform, Abilene, KS.

Obama, B. (2007). *Homeland security*. http://www.whitehouse.gov/agenda/homeland_security/.

Techopedia. (2017). Enterprise Network. Accessed 04/10/17. https://www.techno pedia.com/defintion/7044/enterprise-network.

4

Communications Sector Security

In the prevention of and preparation for terrorism, ending well is the best revenge.

—F.R. Spellman

A terrorist act is a Black Swan.
Why?
The event is a surprise (9/11 attack).
The event has a major effect (changed the world).

After the first recorded instance of the event, it is rationalized by hindsight, as if it could have been expected (recall statements by radical know it all pundits and snowflakes); that is, the relevant data were available but unaccounted for in risk mitigations programs (really, are you serious?) [Seriously adapted and severely modified from Taleb, 2007].

> The emergence of amorphous and largely unknown terrorist individuals and groups operating independently (freelancers) and the new recruitment patterns of some groups, such as recruiting suicide commandos, female and child terrorists, and scientists capable of developing weapons of mass destruction, provide a measure of urgency to increasing our understanding of the psychological and sociological dynamics of terrorist groups and individuals (Hudson, 1999).

The only truly secure system is one that is powered off, cast in a block of concrete and sealed in a lead-lined room with armed guards.

—Gene Spafford

The flow of providing security protection, from data to understanding:

DATA ⟶ INFORMATION ⟶ KNOWLEDGE ⟶ UNDERSTANDING

MR. GRASSHOPPER AND MR. RABBIT #1

In a recent grasshopper-rabbit conversation, a passerby wolf overheard the following:

> "Mr. Rabbit," said grasshopper, "Do you think them terrorists are trying to do away with all them human friends of ours by shutting down their simplistic means to communicate?"
>
> After a moment of deep thought, reflection and a steady but subtle thumping of one of his back feet, Mr. Rabbit replied: "That, my long-legged friend, is a good question . . . and I am not sure what the terrorists are up to these days . . . but one thing I know for certain: when one does not have anything to say in the first place, one has nothing to worry about . . . at least, in that regard."
>
> Grasshopper: "Well, my furry friend, that makes good sense."
>
> Rabbit replied: "Right on, grasshopper!"
>
> Grasshopper nodded in approval of Mr. Rabbit's statement. In the background, the wolf stalked silently back into the forest . . . hidden from view, for the moment. It was obvious to him that Grasshopper and Mr. Rabbit were too wise for him to mess with . . . so he didn't.

INTRODUCTION[1]

From several terrorism incidents on communication sector critical infrastructure in the United States, South America, the Middle East and Africa, it is apparent that communications infrastructure is a preferred target of the

1. The information contained in this chapter is based on DHS (2007) Defense Industrial Base Sector—Sector Plan. http://www.dhs.gov/xlibrary/assets/nipp-spp-defense.

terrorists. Moreover, Al-Qaeda, ISIS, homegrown terrorists and others have made it clear via documented actions or threats that it has an interest in attacking the American communications sector.

The obvious question might be what makes the communications sector such an attractive terrorism target? The communications sector is an attractive target because of the following:

- The communication sector is a particularly attractive target because it is easy to infiltrate and like other lifeline functions—which include energy, transportation, and water, and other resources essential to the operations of most critical infrastructure partners and communities' communications—it is a target of choice.
- The communication sector components or assets are spread throughout the nation with little definition of boundaries.
- Much of the communications sector is equipment that is one-of-a-kind technical electronics (satellites) and antennas which are difficult to replace in the short term.
- Several communication sector components were designed and constructed without concern for terrorist intrusion or destructive activities.
- Many communications systems are monitored and operated using underprotected SCADA systems.
- As with many of industries attempting to economize, many communications segments and sub-segments assign responsibility for safety and security as a collateral duty to a line employee instead of employing a fulltime certified cyber and IT safety and security professional.

COMMUNICATIONS SECTOR SECURITY GOALS AND ATTRIBUTES

The Department of Homeland Security has identified eight general elements and characteristics of critical infrastructure (including communications sector) security goals and attributes.

- **Critical Asset Reduction Goal.** Sector resiliency will be most assured if no particular asset can be assessed as more critical than any other. While the ultimate ideal goal would be zero-critical communication assets, the sector will strive to reduce the number of critical assets whenever and wherever possible within fiscal and legal constraints. Sound risk management practices, including asset resiliency, mitigation of risks and redundancy, will be shared and advanced throughout the sector.

- **Cyber Goals and Attributes.** Like physical attributes, these assist the Communications Sector to evaluate consequences and vulnerabilities, and develop protective strategies. Cyber systems that link and help monitor and control the communications systems are increasingly recognized as a potential vulnerability. All information that identifies or otherwise describes characteristics of a critical communications asset that is created, held, and maintained by the government or the private sector will be protected from unauthorized disclosure according to established procedures, appropriate to the particular level of information.
- **Volumetric or Throughput Attributes.** These define the extent of any damage, depending on the utilized capacity of the systems, or points where the system may be capacity constrained.
- **Personnel Security Goals and Human Attributes.** All personnel directly associated with a critical communication asset are vetted for employment suitability, reliability, and trustworthiness using established processes commensurate with requirements of the respective positions held, in conformance with pertinent security policy. Highly trained and skilled personnel are key factors in a comprehensive Communications Sector security plan. The availability of skilled and experienced technical talent is a concern in the Communications Sector. Sustaining essential technical knowledge is critical to maintaining the sector's safety, reliability, and security.
- **Physical Security Goal.** Determining the impact or consequence of critical communication asset loss and the mission(s) the asset supported, the known or perceived threat, and the susceptibility to exploitation of vulnerabilities the threat is capable of perpetuating is important; identifying specific communications assets the destruction or disruption of which could result in human casualties or economic disruption similar to the effects of weapons of mass destruction is vital; compiling a composite of facility physical security risk assessments is necessary.
- **Insider Threat Goal and Attributes.** Responsible parties in charge provide security education and training aids to communication asset owners/operators not having security programs so that they may implement provisions for the vetting of system and network administrators commensurate with the consequences of the loss of sensitive or classified information, production or provisioning capability, and supply chain integrity.

- **Monitoring and Reporting Goals and Attributes.** Ongoing determination of the effectiveness of government threat reporting to officials, owners, and operators responsible for critical communications assets, and to local law enforcement officials and other first responders including, as appropriate, the medical and mass transportation communities is needed.
- **Training and Education Goal and Attributes.** Developing and providing continuous specific security education and training materials for critical communications asset owners/operators is vital.

HOMELAND SECURITY DIRECTIVES

As a result of 9/11, the Homeland Security Department was formed. On matters pertaining to Homeland Security, Homeland Security Presidential Directives are issued by the president. Each directive has specific meaning and purpose and is carried out by the U.S. Department of Homeland Security. Table 4.1 lists Homeland Security Presidential Directives (HSPDs).

ASSESSING CONSEQUENCES

The potential physical and cyber consequences of any incident, including terror attacks and natural or human-made disasters, are the primary consideration in risk assessment. In the context of this text, consequence is measured as the range of loss or damage that can be expected. The consequences that are considered for the national-level comparative risk assessment are based on the criteria as set forth in HSPD-7. These criteria can be divided into four main categories:

- **Human Impact:** Effect on human life and physical well-being (e.g., fatalities, injures).
- **Economic Impact:** Direct and indirect effects on the economy (e.g., costs resulting from disruption of products or services, costs to respond to and recover from the disruption, costs to rebuild the asset, and long-term costs due to environmental damage).
- **Impact on Public Confidence:** Effect on public morale and confidence in national economic and political institutions.
- **Impact on Government Capability:** Effect on the government's ability to maintain order, deliver minimum essential public services, ensure public health and safety, and carry out national security-related missions.

Table 4.1. Homeland Security Presidential Directives

HSPD – 1: Organization and Operation of the Homeland Security Council. (White House) Ensures coordination of all homeland security-related activities among executive departments and agencies and promotes the effective development and implementation of all homeland security policies.

HSPD – 2: Combating Terrorism through Immigration Policies. (White House) Provides for the creation of task force which will work aggressively to prevent aliens who engage in or support terrorist activity from entering the United States and to detain, prosecute, or deport any such aliens who are within the United States.

HSPD – 3: Homeland Security Advisory System. (White House) Establishes a comprehensive and effective means to disseminate information regarding the risk of terrorist acts to federal, state, and local authorities and to the American people.

HSPD – 4: National Strategy to Combat Weapons of Mass Destruction. Applies new technologies, increases emphasis on intelligence collection and analysis, strengthens alliance relationships, and establishes new partnerships with former adversaries to counter this threat in all of its dimensions.

HSPD – 5: Management of Domestic Incidents. (White House) Enhances the ability of the United States to manage domestic incidents by establishing a single, comprehensive national incident management system.

HSPD – 6: Integration and Use of Screening Information. (White House) Provides for the establishment of the Terrorist Threat Integration Center.

HSPD – 7: Critical Infrastructure Identification, Prioritization, and Protection. (White House) Establishes a national policy for federal departments and agencies to identify and prioritize United States critical infrastructure and key resources and to protect them from terrorist attacks.

HSPD – 8: National Preparedness. (White House) Identifies steps for improved coordination in response to incidents. This directive describes the way federal departments and agencies will prepare for such a response, including prevention activities during the early stages of a terrorism incident. This directive is a companion to HSPD-5.

HSPD – 8 Annex 1: National Planning. Further enhances the preparedness of the United States by formally establishing a standard and comprehensive approach to national planning.

HSPD – 9: Defense of United States Agriculture and Food. (White House) Establishes a national policy to defend the agriculture and food system against terrorist attacks, major disasters, and other emergencies.

HSPD – 10: Biodefense for the 21st Century. (White House) Provides a comprehensive framework for our nation's biodefense.

HSPD – 11: Comprehensive Terrorist-Related Screening Procedures. (White House) Implements a coordinated and comprehensive approach to terrorist-related screening that supports homeland security, at home and abroad. This directive builds upon HSPD – 6.

HSPD – 12: Policy for a Common Identification Standard for Federal Employees and Contractors. (White House) Establishes a mandatory, government-wide standard for secure and reliable forms of identification issued by the federal government to its employees and contractors (including contractor employees).

HSPD – 13: Maritime Security Policy. Establishes policy guidelines to enhance national and homeland security by protecting U.S. maritime interests.

HSPD – 14: Domestic Nuclear Detection establishes a Domestic Nuclear Detection Office (DNSO) to coordinate efforts to protect the domestic United States against dangers from nuclear or radiological materials. EPA supports the detection, response, law enforcement, and information sharing aspects of the DNDO's mission.

HSPD – 16: Aviation Strategy. Details a strategic vision or aviation security while recognizing ongoing efforts, and directs the production of a National Strategy for Aviation Security and supporting plans.

HSPD – 18: Medical Countermeasures against Weapons of Mass Destruction. (White House) Establishes policy guidelines to draw upon the considerable potential of the scientific community in the public and private sectors to address medical countermeasure requirements relating to CBRN threats.

HSPD – 19: Combating Terrorist Use of Explosives in the United States. (White House) Establishes a national policy, and calls for the development of a national strategy and implementation plan, on the prevention and detection of, protection against, and response to terrorist use of explosives in the United States.

HSPD – 20: National Continuity Policy. (White House) Establishes a comprehensive national policy on the continuity of federal government structures and operations and a single National Continuity Coordinator responsible for coordinating the development and implementation of federal continuity policies.

HSPD – 21: Public Health and Medical Preparedness. (White House) Establishes a national strategy that will enable a level of public health and medical preparedness sufficient to address a range of possible disasters.

HSPD – 23: Cybersecurity requires federal agencies to monitor cyber activity toward federal agencies' computer systems and where necessary, provide action to eliminate sources of hostile action. EPA has a robust security program for both personnel and cybersecurity as mandate by the directive.

Source: USEPA (2016).

DID YOU KNOW?

The Department of Homeland Security (DHS) has the mission to provide a common baseline of security across the federal civilian executive branch and to help agencies manage their cyber risk. The common baseline is provided in part through the EINSTEIN system. EINSTEIN serves two key roles in federal government cybersecurity. First, EINSTEIN detects and blocks cyber attacks from compromising federal agencies. Second, EINSTEIN provides DHS with the situational awareness to use threat information detected in one agency to protect the rest of the government and to help the private sector protect itself (DHS, 2017b).

Moreover, HSPD-7 is important to the communications sector in that it required the Department of Homeland Security to "serve as the focal point for the security of cyberspace" with a mission that included "analysis, warning, information sharing, vulnerability reduction, mitigation, and aiding national recovery efforts for critical infrastructure information systems." This directive established a national policy for federal departments and agencies to indentify and prioritize United States critical infrastructure and key resources and to protect them from terrorist attacks. In addition, it required heads of all federal agencies to "develop . . . plans for protecting the physical and cyber critical infrastructure and key resources that they own or operate." Hence, the federal government began to directly address issues of cybersecurity within the federal government systems (FCC, 2017).

As a result of HSPD-7, the Department of Homeland Security established the National Cybersecurity Division (NCSD). The objectives of this division are "to build and maintain an effective national cyberspace response system, and to implement a cyber risk management program for protection of critical infrastructure." The primary operational arms of the division are, first, the Cybersecurity Preparedness and National Cyber Alert System and, secondly, the US Computer Emergency Response Team (US-CERT). The National Cyber Alert System was created by US-CERT and the Department of Homeland Security to help protect computers. One of US-CERT's overarching goals is to ensure that individuals and agencies have access to timely information through tips and alerts about security topics and events. The US-CERT has become the national first line of defense in the war of cybersecurity. The CERT's Cyber Risk Management Program assesses risk, prioritizes resources, and executes protective measures in order to secure the cyber infrastructure. It includes such things as current risk assessments and vulnerabilities that are maintained in their vulnerability database, the National Cyber Alert System, for information dissemination and a number of other references for cybersecurity measures.

In addition to the importance of HSPD-7 providing guidance and direction in cybersecurity and communications protection objectives, as a further shot in the security arm, so to speak, HSPD-23 was signed in January 2008 by President Bush; this directive was necessary due to increased cyber activity on an international scale and attacks targeted at U.S. computers and networks—including computer-controlled systems. HSPD-7 established a Comprehensive National Cybersecurity Initiative (CNCI). Although the document is classified, public sources have indicated that in addition to establishing the National Cybersecurity Center within the Department of Homeland Security, the initiative had twelve other objectives (FCC, 2017):

- Move toward managing a single federal enterprise network
- Deploy intrinsic detection systems
- Develop and deploy intrusion prevention tools
- Review and potentially redirect research and funding
- Connect current government cyber operations centers
- Develop a government-wide cyber intelligence plan
- Increase the security of classified networks
- Expand cyber education
- Define enduring leap-ahead technologies
- Define enduring deterrent technologies and programs
- Develop multi-pronged approaches to supply chain risk management
- Define the role of cybersecurity in private sector domains

DID YOU KNOW?

The federal enterprise network depends on information technology (IT) systems and computer networks for essential operations. These systems face large and diverse cyber threats that range from unsophisticated hackers to technically competent intruders using state-of-the-art intrusion techniques. Many malicious attacks are designed to steal information and disrupt, deny access to, degrade, or destroy critical information systems (DHS, 2017a).

REFERENCES AND RECOMMENDED READING

Crayton, J. W. (1983). Terrorism and the Psychology of the Self, pages 33–41 in Lawrence Zelic Freedman and Yonah Alexander, eds., *Perspectives on Terrorism*. Wilmington, DE: Scholarly Resources.

DHS. (2003). The National Strategy for the Physical Protection of Critical Infrastructures and Key Assets. https://www.dhs.gov/xlibrary/assets/physical_Strat.

DHS. (2010). Communications Sector-Specific Plan. Washington, DC: U.S. Department of Homeland Security.

DHS. (2017a). *Securing Federal Networks*. Accessed April 15, 2017. https:/www.dhs.gov/topic/securing-federal-networks.

DHS. (2017b). *EINSTEIN*. Accessed April 14, 2017. https://www.dhs.gov/einstein.

FCC. (2017). Public Safety Tech Topic #20—Cyber Security and Communications. Accessed 14 April 2017. https://www.fcc.gov/help/public-safety-tech-topic-20-cyber-security-and-communiicaitons.

FEMA. (2015). *Protecting Critical Infrastructure Against Insider Threats*. Accessed 17 April 2015. http://emilms.fema.gov/IS0915/IABsummary.htm.

Ferracuti, F. (1982). A Sociopsychiatric Interpretation of Terrorism. *The Annals of the American Academy of Political and Social Science* 463 (September): 129–41.

FR. (2007). *Federal Register* 17688-17745.

Hudson, R. A. (1999). *The Sociology and Psychology of Terrorism: Who Becomes a Terrorist and Why*? Washington, DC: Library of Congress.

Lees, Frank. (1996). *Loss Prevention in the Process Industries* 3:A5.1–A5.11. New York: Butterworth-Heinemann.

Long, D. E. (1990). *The Anatomy of Terrorism*. New York: Free Press.

Margolin, J. (1977). Psychological Perspectives on Terrorism. In Y. Alexander and S. M. Finger (eds.), *Terrorism: Interdisciplinary Perspectives*. New York: John Jay Press.

Olson, M. (1971). *The Logic of Collective Action*. Boston: Harvard University Press.

OMB. (1998). *Federal Conformity Assessment Activities, Circular A-119*. Washington, DC: White House.

Pearlstein, R. (1991). *The Mind of the Political Terrorist*. Wilmington, DE: Scholarly Resources, Inc.

Sullivant, J. (2007). *Strategies for Protecting National Critical Infrastructure Assets: A Focus on Problem-Solving*. New York: Wiley & Sons.

USEPA. (2016). *Homeland Security Presidential Directives*. https://www.epa.gov/emeergency-resposne/horelamnd -security-presidential-directives.

Wilkinson, P. (1974). *Political Terrorism*. London: Macmillan.

5

Vulnerability Assessment (VA)

In God we trust, all others we monitor.

—Intercept Operator's motto

Vulnerability means different things to different people . . . many associate vulnerability with a specific set of human activities.

—Foster, 1987

But there's one thing we must all be clear about: terrorism is not the pursuit of legitimate goals by some sort of legitimate means. Whatever the murderers may be trying to achieve, creating a better world certainly isn't one of their goals. Instead they are out to murder innocent people.

—Salman Rushdie

INTRODUCTION

One consequence of the events of 9/11 was Department of Homeland Security's (DHS) directive to establish a Critical Infrastructure Protection Task Force to ensure that activities to protect and secure vital infrastructure are comprehensive and carried out expeditiously. Another consequence is a heightened concern among citizens in the United States over the security of their energy infrastructure (i.e., the uninterrupted supply of electrical power and fuel to power vehicles, homes, and vital communications systems). As mentioned, along with other critical infrastructure, the communications sector is classified as "vulnerable" in the sense that inherent weaknesses in its operating environment could be exploited to cause harm to the system. There is also the possibility of a cascading effect—a chain of events—due to a terrorist act affecting communication sector providers, which could cause corresponding damage (collateral damage) to other nearby users. In addition to significant damage to the nation's communications sector, using and needing communications services to function can result in loss of life due to a lack of proper emergency response; shutdown of other industries; loss of electronic communication operational control networks causing catastrophic environmental damage to rivers, lakes and wetlands; and other long-term public health impacts.

ASSESSING VULNERABILITIES

For the purpose of this text and according to FEMA 2008, vulnerability is defined as any weakness that can be exploited by an aggressor to make an asset susceptible to hazard damage.

In addition, according to the Department of Homeland Security (DHS 2009), vulnerabilities are physical features or operational attributes that render an entity open to exploitation or susceptible to a given hazard. Vulnerabilities may be associated with physical (e.g., a broken fence), cyber (e.g., lack of a firewall), or human (e.g., untrained guards) factors.

Vulnerability assessments estimate the odds that a characteristic of, or flaw in, an infrastructure could make it susceptible to destruction, disruption, or exploitation based on its design, location, security posture, processes, or operations. Vulnerabilities typically are identified through internal assessments and information sharing with customers, vendors, and suppliers.

A vulnerability assessment methodology was developed as part of the complete Communications Sector-Specific Plan (CSSP) risk assessment methodology. The methodology examined physical, cyber, and human vulnerabilities and considered relevant national preparedness threat scenarios. The process varied depending on the architecture elements being studied and included subject matter expert interviews, site visits, and modeling and analysis.

The vulnerabilities of communications architecture elements may vary depending on whether they are operational or implementation specific.

Operational vulnerabilities may include those that result from the inherent principles of network design, unanticipated network congestion caused by external factors, or collateral consequences from major disasters or events. Implementation-specific vulnerabilities may be very particular in nature—from bugs in application software and protocol deficiencies to backdoors in vendor equipment firmware or software. The magnitude of the implementation vulnerabilities also varies depending on the exposure of the vulnerable equipment. While embedded firmware, for example, may have only limited exposures to configuration and maintenance functions, systems such as the Domain Name Service require a high degree of exposure in order to provide service to customers (DHS, 2010).

Vulnerability assessments are conducted on many levels. DHS has instituted a process to provide Awareness Training to communication asset owners/operators. The purpose of the Awareness Training is to provide communication sector personnel with information about the place of their asset within the overall communication mission requirements and acquisition process so they will understand their rules and importance to the entities at the corporate and site levels (DHS, 2010). The training focuses on:

- Protection of communication interests
- Protection of Federal interests
- Importance of facilities fostering relationships with local responders and federal, state, and local law enforcement/civil authorities for business recovery planning.

The Awareness Training also informs the asset owners/operators of the protection measures applied to their proprietary and business-sensitive information provided by and to the communication sector. Once critical communications assets are identified and prioritized, the next step is to conduct standardized assessments. DHS, working through and with various agencies, has established a standardized mission assurance assessment for application to critical communication assets. These assessments consider impact, vulnerability, and threat/hazard (whether from natural disaster, technological failure, human error, criminal activity, or terrorist attack). This approach to risk assessment ensures consideration of relevant factors for each communication asset and a relative prioritization of risks to support military operations.

MR. GRASSHOPPER AND MR. RABBIT CONVERSATION #2

Mr. Grasshopper stood guard in the woods while his buddy, Mr. Rabbit, a few feet to his right, gobbled up some succulent plants. A few feet farther to their right, they could hear backpackers talking as they hiked the trail near them. Mr. Grasshopper and Mr. Rabbit, while Mr. Rabbit munched away, hidden from view, listened to the hikers as they came closer, then walked by, and then became silent with the distance.

While the hikers were near, Mr. Grasshopper and Mr. Rabbit were tuned in to the words of the hikers that floated on the air as they trekked along the pathway.

A human male voice said, "I am worried about that insider cyber-hacking and terrorism that is going on at work and making our lives at work harder and miserable . . . very." After a short pause a female voice chimed in and said "Yes, it is hard to get the work done when everything we do seems to leak out to the competition . . . we have already lost a dozen or more co-workers from the slack-off in business . . . the business that is being stolen from us." In almost a chiming rhythm could be heard other voices say as they faded off "Yes, we are being put out of business . . . yes, yes . . . but what are we to do?" Another female voice said, "Well, we need to get rid of those new programs . . . and that cyber hacking . . . we need a firewall . . . or something." And then muted sounds as they passed along the trail and deeper into the woods and then only the munching on succulent plants by Mr. Rabbit and the wind passing high in the pine boughs could be heard.

Mr. Grasshopper asked, "Mr. Rabbit, what did you think of that nonsense them humans were speaking of?"

Mr. Rabbit munched away and then paused, thumped a foot, then looked at Mr. Grasshopper, and said, "When will those humans learn the truth?"

Mr. Grasshopper, surprised, asked, "the truth?"

Putting aside a succulent munch, Mr. Rabbit replied, "Yes, them humans we just heard are obviously computer geeks but they seem not to understand that corrupted computer files, programs, digital networks and in-house terrorist acts are not the problem with all them or their computers . . . them insider threat programs and methods they spoke of miss the point."

"Miss the point?" Mr. Grasshopper stated.

"Yes, for sure . . . you see, insider threat programs are not technology-focused and centered on the digital cyber world and its networks. No way,

Jose and Maria . . . you see, Mr. Grasshopper, the insider threat is not like you and me . . . no way, Jose and Maria . . . it is a human, a person, a mammal, Homo sapiens, a breathing organism, a heartbeat."

"Wow . . . I get it!" said Mr. Grasshopper.

"I knew you would," said Mr. Rabbit.

"You are just sooooooooooooooooooooo smart, Mr. Rabbit."

"Right on, Mr. Grasshopper."

INSIDER THREAT VULNERABILITY

Some readers may view the previous discussion between Mr. Grasshopper and Mr. Rabbit as silly, trivial, child's play, ridiculous or maybe not suitable for the discussion contained in this text. However, the previous discussion was not intended to focus on the Rabbit and Grasshopper; instead, the focus was on presenting a true representation of what insider threat is all about. The insider threat to an organization's communication network and cyber activities is not about rocket science; instead it is about malicious software (malware) and the variety of forms of hostile or intrusive software, including computer viruses, worms, Trojan horses, ransomware, spyware, adware, scareware and others. One does need not be a super genius to understand what insider threat really is. To quote an expert, Mr. Rabbit said it best: "the insider threat is . . . a human, a person, a mammal, Homo sapiens, a breathing organism, a heartbeat." The point is that to protect an organization from insider threats the organization must "use tools to monitor the traffic in or out of the networks and be able to focus that monitoring on specific people who do something questionable or suspicious, moreover, nontechnical employee behavior must be monitored" (INSA, 2017).

"Behavior must be monitored" . . . yes, for sure. What monitoring really comes down to, however, is awareness. An important part of any successful vulnerability assessment process is awareness. Security and risk managers (and all employees in general) working in or with the critical infrastructure sectors must be aware of the potential for insider threat vulnerability. Again, the key word is AWARENESS. Awareness means that personnel within the critical infrastructure sectors must know how to identify and take action against insider threats. To achieve this critical goal, safety and security personnel must be provided with an overview of and be cognizant of common characteristics and indicators associated with malicious insiders and effective measures to counter insider threats.

Protecting Against Insider Threats (FEMA, 2015)

As mentioned earlier, when analyzing threats to our nation's critical infrastructure, we tend to focus on malicious actions from boat or plane loads of outside actors. Of equal concern (and even more so in the author's view) are threats from an insider—someone we have given legitimate access to information, systems, and resources. The measures we take to detect and protect against external threats may not be sufficient to address threats from insiders.

A malicious insider has access and inside knowledge of the organization and uses that knowledge with the intent to cause harm. The insider may be a current employee, a former employee, a service provider, or, especially in the current era, a planted person inside who has been brainwashed and turned into a terrorist waiting for the moment of maximum impact to people and property.

Given the importance of our nation's critical infrastructure, the actions taken by a malicious employee or service provider could have devastating consequences.

Let's look at some actual examples.

- A service provider employee at a nuclear facility stole two five-gallon containers of low-enriched uranium dioxide and then attempted to extort $100,000 by threatening to disperse the material in an unnamed U.S. city.
- A power company field engineer, angry with his supervisor, disabled protection systems at a substation and forced the shutdown of the entire network. More than 800,000 customers lost power as a result.
- Two municipal employees used their access credentials to sabotage the system controlling the traffic lights of a major city, causing widespread traffic delays. The damage took four days to repair.
- A disgruntled supermarket meat packaging employee intentionally contaminated hamburger meat with a pesticide, causing various levels of illness in 92 consumers.

Insider threats endanger the integrity and security of our workplaces and our communities. This section helps you become aware of threat indicators and actions you can take. As stated earlier, and it can't be overstated, awareness is the first step to keeping our nation and workplaces safe.

Insider Threat Defined

The President's National Infrastructure Advisory Council (2008) defines the insider threat as follows:

> The insider threat to critical infrastructure is one or more individuals with the access or inside knowledge of a company, organization, or enterprise that would allow them to exploit the vulnerabilities of the entity's security, systems, services, products, or facilities with the intent to cause harm.

A person who takes advantage of access or inside knowledge in such a manner commonly is referred to as a "malicious insider."

The Scope of Insider Threats

Insider threats can be accomplished through either physical or cyber means and may involve any of the following:

- **Physical or information-technology sabotage**—Involves modification or damage to an organization's facilities, property, assets, inventory, or systems with the purpose of harming or threatening harm to an individual, the organization, or the organization's operations.

DID YOU KNOW?

The FBI testified in June 2012 that in the preceding year, economic espionage losses to the American economy totaled more than $13 billion. In the previous four years, the number of arrests the FBI made had doubled; indictments increased by a factor of five; and convictions increased by a factor of eight (FBI, 2012). In another survey, security professionals found that 43.2% of respondents attributed some loss at their organization to insiders, and 46% of respondents said the damage caused by insider attacks was more damaging than outside attacks (CSI, 2011).

- **Theft of intellectual property**—Involves removal or transfer of an organization's intellectual property outside the organization through physical or electronic means (also known as economic espionage).
- **Theft or economic fraud**—Involves acquisition of an organization's financial or other assets through theft or fraud.
- **National security espionage**—Involves obtaining information or assets with a potential impact on national security through clandestine activities.

Common Characteristics and Traits of Malicious Insiders

Based on research conducted by the Software Engineering Institute at Carnegie Mellon University and the U.S. Secret Service National Threat Assessment Center, malicious insiders often are perceived or known to be difficult or high-maintenance employees who are:

- Obviously unhappy or extremely resentful;
- Having financial, performance, or behavioral problems;
- At risk (or perceived to be) for layoff or termination.

Keep in mind that not all malicious insiders fit this characterization. Insiders involved in national security espionage, for example, may exhibit few outward signs. In the majority of cases, however, management and/or human resources personnel were well aware of the employees and their issues prior to an incident.

Personal Factors Associated with Insiders

The following motives and personal situations frequently are linked with malicious insiders:

- **Personal or Behavioral Problems**
  - Vulnerable to blackmail
  - Experiencing family or financial problems
  - Prone to compulsive or destructive behavior
  - Subject to ego or self-image issues

- **Personal Desires**
    - Seeking adventure or thrill
    - Seeking approval and returned favors
    - Professing allegiance
- **Workplace Issues**
    - Experiencing problems at work
    - Feeling anger or need for revenge

Organizational Factors That Embolden Malicious Insiders

The following organizational factors have been known to encourage or present opportunities to potential malicious insiders.

- **Access and Availability**
    - Ease of access to materials and information
    - Ability to exit the facility or network with materials or information
- **Policies and Procedures**
    - Undefined or inadequate policies and procedures
    - Inadequate training
    - Lack of training
- **Time Pressure and Consequences**
    - Rushed employees
    - Perception of lack of consequences

Insider Activities and Behavior You May See

Insider threats may be detected through particular activities and behavior on the part of the insider. This section of the presentation identifies those indicators. These activities and behaviors often will appear unusual or suspicious. Keep in mind there may be several explanations for a particular activity or behavior identified here, but when combined with other factors, certain activities or behaviors may point toward a possible insider threat. A combination or confluence of indicators should not be ignored.

Types of Insider Activities and Behavior

Unusual or suspicious insider activities and behavior can be described using the following categories:

- Inappropriate Interest or Acquisition
- Unauthorized or Unusual Computer Use
- Unusual Hours, Contacts, or Travel
- Secretive or Peculiar Behavior
- Personal or Financial Issues

Employer Actions

- Clearly communicating and consistently enforcing security policies and controls.
- Ensuring that proprietary information and materials are adequately, if not robustly, protected.
- Routinely monitoring computer networks for suspicious activity.
- Ensuring security (to include computer network security) personnel have the tools they need.
- Consulting with legal and law enforcement experts as needed to ensure compliance with the law.

Employee Actions

Critical infrastructure organizations today employ a number of security measures to reduce the risk of insider threats. The measures involving employees include, but are not limited to:

- Using appropriate screening processes to select new employees.
- Educating employees about security or other protocols.
- Encouraging and providing non-threatening, convenient ways for employees to report suspicions in a confidential manner.
- Becoming familiar with behavior and activities associated with malicious insiders.
- Documenting and evaluating incidents of suspicious or disruptive behavior.
- Consulting with legal and law enforcement experts as needed to ensure compliance with the law.

VULNERABILITY ASSESSMENT (VA)[1]

A *vulnerability assessment* involves an in-depth analysis of the facility's functions, systems, and site characteristics to identify facility weaknesses and lack of redundancy, and determine mitigations or corrective actions that can be designed and implemented to reduce the vulnerabilities. A vulnerability assessment can be a stand-alone process or part of a full risk assessment. During this assessment, the analysis of site assets is based on a) the identified threat; b) the criticality of the assets; and c) the level of protection chosen (i.e., based on willingness or unwillingness to accept risk).

It is important to point out that post-9/11 all sectors have taken great strides to protect their critical infrastructure. For instance, government and industry have developed vulnerability assessment methodologies for critical infrastructure systems and trained thousands of auditors and others to conduct them.

The actual complexity of vulnerability assessments will range based upon the design and operation of the communication asset. The nature and extent of the VA will differ among systems based on a number of factors, including system size and potential population, and safety evaluations also vary based on knowledge and types of threats, available security technologies, and applicable local, state and federal regulations. Preferably, a VA is "performance-based," meaning that it evaluates the risk to the communication assets based on the effectiveness (performance) of existing and planned measures to counteract adversarial actions. According to USEPA (2002), the common elements of Communication Sector vulnerability assessments are as follows:

- Characterization of the Communication Sector, including its mission and objectives.
- Identification and prioritization of adverse consequences to avoid.
- Determination of critical assets that might be subject to malevolent acts that could result in undesired consequences.
- Assessment of the likelihood (qualitative probability) of such malevolent acts from adversaries.

1. Much of the information in this section is from US Department of Energy (DOE 2002) *Vulnerability Assessment Methodology: Electric Power Infrastructure*. Washington, DC; US Army Research Laboratory (2000) *Vulnerability Risk Assessment, ARL-TR-1045*. Washington, DC: DOD; U.S. Dept of Justice, (2002) *A Method to Assess the Vulnerability of U.S. Chemical Facilities*. Washington, DC.

- Evaluation of existing countermeasures.
- Analysis of current risk and development of a prioritized plan for risk reduction.

Benefits of Assessments

Communication Sector members should routinely perform vulnerability assessments to better understand threats and vulnerabilities, determine acceptable levels of risk, and stimulate action to mitigate identified vulnerabilities. The direct benefits of performing a vulnerability assessment include:

- **Build and broaden awareness**—the assessment process directs senior management's attention to security. Security issues, risks, vulnerabilities, mitigation options, and best practices are brought to the surface. Awareness is one of the least expensive and most effective methods for improving the organization's overall security posture.
- **Establish or evaluate against a baseline**—if a baseline has been previously established, an assessment is an opportunity for a checkup to gauge the improvement or deterioration of an organization's security posture. If no previous baseline has been set (or the work was not uniform or comprehensive), an assessment is an opportunity to integrate and unify previous efforts, define common metrics, and establish a definitive baseline. The baseline also can be compared against best practices to provide perspective on an organization's security posture.
- **Identify vulnerabilities and develop responses**—generating lists of vulnerabilities and potential responses is usually a core activity and outcome of an assessment. Sometimes, due to budget, time, complexity, and risk considerations, the response selected for many of the vulnerabilities may be non-action, but after completing the assessment process these decisions will be conscious ones, with a documented decision process and item-by-item rationale available for revisiting issues at scheduled intervals. This information can help drive or motivate the development of a risk management process.
- **Categorize key assets and drive the risk management process**—an assessment can be a vehicle for reaching corporate-wide consensus on a hierarchy of key assets. This ranking, combined with threat, vulnerability, and risk

analysis, is at the heart of any risk management process. For many organizations, Y2K was the first time a company-wide inventory and ranking of key assets was attempted. An assessment allows any organization to revisit that list from a broader and more comprehensive perspective.

- **Develop and build internal skills and expertise**—a security assessment, when not implemented in an "audit" mode, can serve as an excellent opportunity to build security skills and expertise within an organization. A well-structured assessment can have elements that serve as a forum for cross-cutting groups to come together and share issues, experiences, and expertise. External assessors can be instructed to emphasize "teaching and collaborating" rather than "evaluating" (the traditional role). Whatever the organization's current level of sophistication, a long-term goal should be to move the organization toward a capability for self-assessment.
- **Promote action**—although disparate security efforts may be underway in an organization, an assessment can crystallize and focus management attention and resources on solving specific and systemic security problems. Often the people in the trenches are well aware of security issues (and even potential solutions) but are unable to convert their awareness to action. An assessment provides an outlet for their concerns and the potential to surface these issues at appropriate levels (legal, financial, executive) and achieve action. A well-designed and executed assessment not only identifies vulnerabilities and makes recommendations; it also gains executive buy-in, identifies key players, and establishes a set of cross-cutting groups that can convert those recommendations into action.
- **Kick off an ongoing security effort**—an assessment can be used as a catalyst to involve people throughout the organization in security issues, build cross-cutting teams, establish permanent forums and councils, and harness the momentum generated by the assessment to build an ongoing institutional security effort. The assessment can lead to the creation of either an actual or a virtual (matrixed) security organization.

Vulnerability Assessment Process

Table 5.1 provides an overview of the elements included in the assessment methodology. The elements included in this overview are based on actual in-field experience and lessons learned.

Table 5.1. Basic Elements in Vulnerability Assessments

Element	*Points to Consider*
1. Characterization of the communications entity, including its mission and objectives.	• What are the important missions of the system to be assessed? Define the highest priority services provided by the utility. Identify the industry's customers: * General public * Government * Military * Industrial * Critical care * Retail operations * Firefighting • What are the most important facilities, processes, and assets of the system for achieving the mission objectives and avoiding undesired consequences? Describe: * Industry facilities * Operating procedures * Management practices that are necessary to achieve the mission objectives * How the industry operates * Treatment processes * Storage methods and capacity * Energy use and storage * Distribution system In assessing those assets that are critical, consider critical customers, dependence on other infrastructures (e.g., chemical, transportation, communications), contractual obligations, single points of failure, chemical hazards and other aspects of the industry's operations, or availability of industry utilities that may increase or decrease the criticality of specific facilities, processes and assets.
2. Identification and prioritization of adverse consequences to avoid.	• Take into account the impacts that could substantially disrupt the ability of the system to provide a safe and reliable supply of materials. Communications sector systems should use the vulnerability assessment process to determine how to reduce risk associated with the consequences of significant concern.

Element	*Points to Consider*
	• Ranges of consequences or impacts for each of these events should be identified and defined. Factors to be considered in assessing the consequences may include: * Magnitude of service disruption * Economic impact (such as replacement and installation costs for damaged critical assets or loss of revenue due to service outage) * Number of illnesses or deaths resulting from an event * Impact on public confidence in the material supply * Chronic problems arising from specific events * Other indicators of the impact of each event as determined by the communications sector Risk reduction recommendations at the conclusion of the vulnerability assessment strive to prevent or reduce each of these consequences.
3. Determination of critical assets that might be subject to malevolent acts that could result in undesired consequences.	• What are the malevolent acts that could reasonably cause undesired consequences? * Electronic, computer, or other automated systems which are utilized by the communications sector entities (e.g., Supervisory Control and Data Acquisition [SCADA]) * The use, storage, or handling of various communication supplies * The operation and maintenance of such systems
4. Assessment of the likelihood (qualitative probability) of such malevolent acts from adversaries.	• Determine the possible modes of attack that might result in consequences of significant concern based on critical assets of the communication sector entity. The objective of this step of the assessment is to move beyond what is merely possible and determine the likelihood of a particular attack scenario. This is a very difficult task as there is often insufficient information to determine the likelihood of a particular event with any degree of certainty.

(*Continued*)

Table 5.1. (Continued)

Element	*Points to Consider*
	• The threats (the kind of adversary and the mode of attack) selected for consideration during a vulnerability assessment will dictate, to a great extent, the risk reduction measures that should be designed to counter the threat(s). Some vulnerability assessment methodologies refer to this as a "Design Basis Threat" (DBT) where the threat serves as the basis for the design of countermeasures, as well as the benchmark against which vulnerabilities are assessed. Differences in geographic location, size of the utility, previous attacks in the local area, and many other factors will influence the threat(s) that the DIB sector entity should consider in the assessments. DIB sector entities should consult with the local FBI and/or other law enforcement agencies, public officials, and others to determine the threats upon which their risk reduction measures should be based. It should be noted that there is no single DBT or threat for all communications systems in the United States. Differences in geographic location, size of the utility, previous attacks in the local area and many other factors will influence the threat(s) that the communications sector entity should consider in the assessments. Communications sector entities should consult with the local FBI and/or other law enforcement agencies, public officials, and others to determine the threats upon which their risk reduction measures should be based.
5. Evaluation of existing countermeasures. (Depending on countermeasures already in place, some critical assets may already be sufficiently protected. This step will aid in identification of the areas of greatest concern and help to focus priorities for risk reduction.)	• What capabilities does the system currently employ for detection, delay, and response? * Identify and evaluate current detection capabilities such as intrusion detection systems, energy quality monitoring, operational alarms, guard post orders, and employee security awareness programs. * Identify current delay mechanisms such as locks and key control, fencing, structure integrity of critical assets, and vehicle access checkpoints.

Element	*Points to Consider*
	* Identify existing policies and procedures for evaluation and response to intrusion and system malfunction alarms, and cybersystem intrusions. **It is important to determine the performance characteristics. Poorly operated and maintained security technologies provide little or no protection.** • What cyber protection system features does the facility have in place? Assess what protective measures are in place for the SCADA and business-related computer information systems, such as: * Firewalls * Modem access * Internet and other external connections, including wireless data and voice communications * Security polices and protocols **It is important to identify whether vendors have access rights and/or "backdoors" to conduct system diagnostics remotely.** • What security policies and procedures exist, and what is the compliance record for them? Identify existing policies and procedures concerning: * Personal security * Physical security * Key and access badge control * Control of system configuration and operational data * Chemical and other vendor deliveries * Security training and exercise records
6. Analysis of current risk and development of a prioritized plan for risk reduction,	• Information gathered on threat, critical assets, communications sector operations, consequences, and existing countermeasures should be analyzed to determine the current level of risk. The utility should then determine whether current risks are acceptable or risk reduction measures should be pursued. • Recommended actions should measurably reduce risks by reducing vulnerabilities and/or consequences through improved deterrence, delay, detection, and/or response capabilities or by improving operational policies or procedures.

(*Continued*)

Table 5.1. (Continued)

Element	*Points to Consider*
	Selection of specific risk reduction actions should be completed prior to considering the cost of the recommended action(s). Facilities should carefully consider both short- and long-term solutions. An analysis of the cost of short- and long-term risk reduction actions may impact which actions the utility chooses to achieve its security goals. • Facilities may also want to consider security improvements. Security and general infrastructure may provide significant multiple benefits. For example, improved treatment processes or system redundancies can both reduce vulnerabilities and enhance day-to-day operation. • Generally, strategies for reducing vulnerabilities fall into three broad categories: * Sound business practices—affect policies, procedures, and training to improve the overall security-related culture at the facility. For example, it is important to ensure rapid communication capabilities exist between public health authorities and local law enforcement and emergency responders. * System upgrades—include changes in operations, equipment, processes, or infrastructure itself that make the system fundamentally safer. * Security upgrades—improve capabilities for detection, delay, or response.

Adapted from Spellman (2009), *The Handbook of Safety Engineering.*

In Table 5.1, step 3 deals with identification of asset criticality. This is an important step in any vulnerability assessment. Identifying asset criticality serves several functions:

- It enables more careful consideration of factors that affect risk, including threats, vulnerabilities, and consequences of loss or compromise of the asset.
- It enables more focused and thorough consideration of loss or compromise of the asset.

- It enables leaders to develop robust methods for managing consequences of asset loss (restoration).
- It provides a means to increase awareness of a broad range of employees to protect truly critical assets and to differentiate in policies and procedures the heightened protection they require.

As previously indicated, identifying the criticality of assets is used primarily to focus the vulnerability analysis efforts. It also assists with the ranking of various recommendations for reducing vulnerabilities. As an example, let's take a look at the criticality of electric power assets and operations included in the normal operation of Communications Sector assets:

Physical

- Generators
- Substations
- Transformers
- Transmission lines
- Distribution lines
- Control center
- Warehouses
- Office buildings
- Internal and external infrastructure dependencies

Cyber

- SCADA systems
- Networks
- Databases
- Business systems
- Telecommunications

Interdependencies

- Single-point nodes of failures
- Critical infrastructure components of high reliance

VULNERABILITY ASSESSMENT METHODOLOGY

Vulnerability assessment methodology consists of ten elements. Each element along with a description of each is listed below (US DOE, 2002).

1. Network architecture
2. Threat environment
3. Penetration testing
4. Physical security
5. Physical asset analysis
6. Operations security
7. Policies and procedures
8. Impact analysis
9. Infrastructure interdependencies
10. Risk characterization

Network Architecture

This element provides an analysis of the information assurance features of the information network(s) associated with the organization's critical information systems. Information examined should include network topology and connectivity (including subnets), principal information assets, interface and communication protocols, function and linage of major software and hardware components (especially those associated with information security such as intrusion detectors), and policies and procedures that govern security features of the network.

Procedures for information assurance in the system, including authentication of access and management of access authorization, should be reviewed. The assessment should identify any obvious concerns related to architectural vulnerabilities, as well as operating procedures. Existing security plans should be evaluated, and the results of any prior testing should be analyzed. Results from the network architecture assessment should include potential recommendations for changes in the information architecture, functional areas and categories where testing is needed, and suggestions regarding system design that would enable more effective information and information system protection.

Three techniques are often used in conducting the network architecture assessment:

1. Analysis of network and system documentation during and after the site visit
2. Interview with facility staff, managers, and chief information officer
3. Tours and physical inspections of key facilities

Threat Environment

Development of a clear understanding of the threat environment is a fundamental element of risk management. When combined with an appreciation of the value of the information assets and systems, and the impact of unauthorized access and subsequent malicious activity, an understanding of threats provides a basis for better defining the level of investment needed to prevent such access.

The threat of a terrorist attack to communications sector infrastructure is real and could come from several areas, including physical, cyber, and interdependency. In addition, threats could come from individuals or organizations motivated by financial gain or persons who derive pleasure from such penetration (e.g., recreational hackers, disgruntled employees). Other possible sources of threats are those who want to accomplish extremist goals (e.g., environmental terrorists, antinuclear advocates) or embarrass one or more organizations.

This element should include a characterization of these and other threats, identification of trends in these threats, and ways in which vulnerabilities are exploited. To the extent possible, characterization of the threat environment should be localized, that is, within the organization's service area.

Penetration Testing

The purpose of network penetration testing is to utilize active scanning and penetration tools to identify vulnerabilities that a determined adversary could easily exploit. Penetration testing can be customized to meet the specific needs and concerns of the communications sector unit. In general, penetration testing should include a test plan and details on the rules of engagement

(ROE). It should also include a general characterization of the access points of the critical information systems and include a general characterization of the access points to the critical information systems and communication interface connections, modem network connections, access points to principal network routers, and other external connections. Finally, penetration testing should include identified vulnerabilities and, in particular, whether access could be gained to the control network or specific subsystem or devices that have a critical role in assuring continuity of service.

Penetration testing consists of an overall process of establishing the ground rules or ROE for the test; establishing a white cell for continuous communication; developing a format or methodology for the test; conducting the test; and generating a final report that details methods, findings, and recommendations.

Penetration testing methodology consists of three phases: reconnaissance, scenario development, and exploitation. A one-time penetration test can provide the utility with valuable feedback; however, it is far more effective if performed on a regular basis. Repeated testing is recommended because new threats develop continuously, and the networks, computers, and architecture of the communications sector unit or utility are likely to change over time.

Physical Security

A critical dependency for the Communications Sector as well as the other sectors is related to the physical security of the facilities. The purpose of physical security assessment is to examine and evaluate the systems in place (or being planned) and to identify potential improvements in this area for the sites evaluated. Physical security systems include access controls, barriers, locks and keys, badges and passes, intrusion detection devices and associated alarm reporting and display, closed-circuit television (assessment and surveillance), communications equipment (telephone, two-way radio, intercom, cellular), lighting (interior and exterior), power sources (line, battery, generator), inventory control, postings (signs), security system wiring, and protective force. Physical security systems are reviewed for design, installation operation, maintenance, and testing.

The physical security assessment should focus on those sites directly related to the critical facilities, including information systems and assets required for operation. Typically included are facilities that house critical equipment or information assets or networks dedicated to the operation of electric, oil, or gas transmission, storage, or delivery systems. Other facilities can be included on the basis of criteria specified by the organization being assessed. Appropriate levels of physical security are contingent upon the value of company assets, the potential threats to these assets, and the cost associated with protecting the assets. Once the cost of implementing/maintaining physical security programs is known, it can be compared to the value of the company assets, thus providing the necessary information for risk management decisions. The focus of the physical security assessment task is determined by prioritizing the company assets; that is, the most critical assets receive the majority of the assessment activity.

At the start of the assessment, survey personnel should develop a prioritized listing of company assets. This list should be discussed with company personnel to identify areas of security strengths and weaknesses. During these initial interviews, assessment areas that would provide the most benefit to the company should be identified; once known, they should become the major focus of the assessment activities.

The physical security assessment of each focus area usually consists of the following:

- Physical security program (general)
- Physical security program (planning)
- Barriers
- Access controls/badges
- Locks/keys
- Intrusion detection systems
- Communications equipment
- Protective force/local law enforcement agency

The key to reviewing the above topics is not to just identify if they exist but to determine the appropriate level that is necessary and consistent with the value of the asset being protected. The physical security assessment worksheets provide guidance on appropriate levels of protection.

Once the focus and content of the assessment task have been identified, the approach to conducting the assessment can be either at the "implementation level" or at the "organizational level." The approach taken depends on the maturity of the security program.

For example, a company with a solid security infrastructure (staffing plans/procedures, funding) should receive a cursory review of these items; however, facilities where the security programs are being implemented should receive a detailed review. The security staff can act upon deficiencies found at the facilities, once reported.

For companies with an insufficient security organization, the majority of time spent on the assessment should take place at the organizational level to identify the appropriate staffing/funding necessary to implement security programs to protect company assets. Research into specific facility deficiencies should be limited to finding just enough examples to support any staffing/funding recommendations.

Physical Asset Analysis

The purpose of the physical asset analysis is to examine the systems and physical operational assets to ascertain whether vulnerabilities exist. Included in this element is an examination of asset utilization, system redundancies, and emergency operating procedures. Consideration should also be given to the topology and operating practices for electric and gas transmission, processing, storage, and delivery, looking specifically for those elements that either singly or in concert with other factors provide a high potential for disrupting service. This portion of the assessment determines company and industry trends regarding these physical assets. Historic trends, such as asset utilization, maintenance, new infrastructure investments, spare parts, SCADA linkages, and field personnel are part of the scoping element.

The proposed methodology for physical assets is based on a macro-level approach. The analysis can be performed with company data, public data, or both. Some companies might not have readily available data or might be reluctant to share that data.

Key output from analysis should be graphs that show trends. The historic data analysis should be supplemented with on-site interviews and visits. Items to focus on during a site visit include the following:

- Trends in field testing
- Trends in maintenance expenditures
- Trends in infrastructure investments
- Historic infrastructure outages

- Critical system components and potential system bottlenecks
- Overall system operation controls
- Use and dependency of SCADA systems
- Linkages of operation staff with physical and IT security
- Adequate policies and procedures
- Communications with other regional utilities
- Communications with external infrastructure providers
- Adequate organizational structure

Operations Security

Operations security (OPSEC) is the systematic process of denying potential adversaries (including competitors or their agents) information about capabilities and intentions of the host organization. OPSEC involves identifying, controlling, and protecting generally nonsensitive activities concerning planning and execution of sensitive activities. The OPSEC assessment reviews the processes and practices employed for denying adversary access to sensitive and nonsensitive information that might inappropriately aid or abet an individual's or organization's disproportionate influence over system operation. This assessment should include a review of security training and awareness programs, discussions with key staff, and tours of appropriate principal facilities. Information that might be available through public access should also be reviewed.

Policies and Procedures

The policies and procedures by which security is administered (1) provide the basis for identifying and resolving issues; (2) establish the standards of reference for policy implementation; and (3) define and communicate roles, responsibilities, authorities, and accountabilities for all individuals and organizations that interface with critical systems. They are the backbone for decisions and day-to-day security operations. Security policies and procedures become particularly important at times when multiple parties must interact to effect a desired level of security and when substantial legal ramifications cold result from policy violations. Policies and procedures should be reviewed to determine whether they (1) address the key factors affecting security; (2) enable effective compliance, implementation, and enforcement; (3) reference or conform to established standards; (4) provide clear and comprehensive guidance; and (5) effectively address the roles, responsibilities, accountabilities, and authorities.

The objective of the policies and procedures assessment task is to develop a comprehensive understanding of how a facility protects its critical assets through the development and implementation of policies and procedures. Understanding and assessing this area provide a means of identifying strengths and areas for improvements that can be achieved through:

- Modification of current policies and procedures
- Implementation of current policies and procedures
- Development and implementation of new policies and procedures
- Assurance of compliance with policies and procedures
- Cancellation of policies and procedures that are no longer relevant, or are inappropriate, for the facility's current strategy and operations

Impact Analysis

A detailed analysis should be conducted to determine the influence that exploitation of unauthorized access to critical facilities or information systems might have on an organization's operations (e.g., market and/or physical operations). In general, such an analysis would require thorough understanding of (1) the applications and their information processing, (2) decisions influenced by this information, (3) independent checks and balances that might exist regarding information upon which decisions are made, (4) factors that might mitigate the impact of unauthorized access, and (5) secondary impacts of such access. Similarly, the physical chain of events following disruption, including the primary, secondary, and tertiary impacts of disruption, should be examined.

The purpose of the impact analysis is to help estimate the impact that detrimental impacts could have on communication sector units. The impact analysis provides an introduction to risk characterization by providing quantitative estimates of these impacts so that the communication sector unit can implement a risk management program and weigh the risks and costs of various mitigation measures.

Infrastructure Interdependencies

The term "infrastructure interdependencies" refers to the physical and electronic (cyber) linkages within communications and among our nation's critical infrastructures—energy (electric power, oil, natural gas), telecommunications, transportation, water supply systems, banking and finance, emergency services, and government services. This task identifies the direct infrastructure linkages between and among the infrastructures that support critical facilities as recognized by the organization. Performance of this task requires a detailed understanding of an organization's functions, internal infrastructures, and how these link to external infrastructures.

The purpose of the infrastructure interdependencies assessment is to examine and evaluate the infrastructures (internal and external) that support critical facility functions, along with their associated interdependencies and vulnerabilities.

Risk Characterization

Risk characterization provides a framework for prioritizing recommendations across all task areas. The recommendations for each task area are judged against a set of criteria to help prioritize the recommendations and assist the organization in determining the appropriate course of action. It provides a framework for assessing vulnerabilities, threats, and potential impacts (determined in the other tasks). In addition, the existing risk analysis and management process at the organization should be reviewed and, if appropriate, utilized for prioritizing recommendations. The degree to which corporate risk management includes security factors is also evaluated.

VULNERABILITY ASSESSMENT PROCEDURES

Vulnerability assessment procedures can be conducted for Communications Sector assets using various methodologies. For example, the checklist analysis is an effective technology. In addition, Pareto analysis (80/20 principle), relative ranking, pre-removal risk assessment (PRRA), change analysis, failure mode and effects analysis (FMEA), fault tree analysis, event tree analysis, what-if analysis and Hazard and Operability (HAZOP) can be used in conducting the assessment.

Based on personal experience, the what-if analysis and HAZOP seem to be the most user-friendly methodologies to use. A sample what-if analysis procedural outline is presented below, followed by a brief explanation and outline for conducting HAZOP.

What-If Analysis Procedure/Sample What-If Questions

The steps in a What-If Checklist analysis are as follows:

1. Select the team (personnel experienced in the process)
2. Assemble information (piping and instrumentation drawings (P &IDs), process flow diagrams (PFDs), operating procedures, equipment drawings, etc.).
3. Develop a list of what-if questions.
4. Assemble your team in a room where each team member can view the information.
5. Ask each what-if question in turn and determine:
    - What can cause the deviation from design intent that is expressed by the question?
    - What adverse consequences might follow?
    - What are the existing design and procedural safeguards?
    - Are these safeguards adequate?
    - If these safeguards are not adequate, what additional safeguards does the team recommend?
6. As the discussion proceeds, record the answers to these questions in tabular format.
7. Do not restrict yourself to the list of questions that you developed before the project started. The team is free to ask additional questions at any time.
8. When you have finished the what-if questions, proceed to examine the checklist. The purpose of this checklist is to ensure that the team has not forgotten anything. While you are reviewing the checklist, other what-if questions may occur to you.
9. Make sure that you follow up all recommendations and action items that arise from the hazards evaluation.

HAZOP Analysis

The HAZOP analysis technique uses a systematic process to (1) identify possible deviations from normal operations and (2) ensure that safeguards are in place to help prevent accidents. The HAZOP uses special adjectives (such as speed, flow, pressure, etc.) combined with process conditions (such as "more," "less," "no," etc.) to systematically consider all credible deviations from normal conditions. The adjectives, called guide words, are a unique feature of HAZOP analysis.

In this approach, each guide word is combined with relevant process parameters and applied at each point (study node, process section, or operating step) in the process that is being examined.

Table 5.2. HAZOP Analysis Guide Words

Guide Words	*Meaning*
No	Negation of the Design Intent
Less	Quantitative Decrease
More	Quantitative Increase
Part Of	Other Material Present by Intent
As Well As	Other Materials Present Unintentionally
Reverse	Logical Opposite of the Intent
Other Than	Complete Substitution

Table 5.3. Common HAZOP Analysis Process Parameters

Flow	*Time*	*Frequency*	*Mixing*
Pressure	Composition	Viscosity	Addition
Temperature	pH	Voltage	Separation
Level	Speed	Information	Reaction

The following is an example of creating deviations using guide words and process parameters:

Guide Words		**Parameter**		**Deviation**
NO	+	FLOW	=	NO FLOW
MORE	+	PRESSURE	=	HIGH PRESSURE
AS WELL AS	+	ONE PHASE	=	TWO PHASE
OTHER THAN	+	OPERATION	=	MAINTENANCE
MORE	+	LEVEL	=	HIGH LEVEL

Guide words are applied to both the more general parameters (e.g., react, mix) and the more specific parameters (e.g., pressure, temperature). With the general parameters, it is not unusual to have more than one deviation from the application of one guide word. For example, "more reaction" could mean either that a reaction takes place at a faster rate, or that a greater quantity of product results. On the other hand, some combination of guide words and parameters will yield no sensible deviation (e.g., "as well as" with "pressure").

HAZOP Procedure

1. Select the team
2. Assemble information (P&IDs, PFDs, operating procedures, equipment drawings, etc.).
3. Assemble your team in a room where each team member can view P&IDs.
4. Divide the system you are reviewing into nodes (you can present the nodes, or the team can choose them as you go along).
5. Apply appropriate deviations to each node. For each deviation, address the following questions:
    - What can cause the deviation from design intent?
    - What adverse consequences might follow?
    - What are the existing design and procedural safeguards?
    - Are these safeguards adequate?
    - If these safeguards are not adequate, what does the team recommend?
6. As the discussion proceeds, record the answers to these questions in tabular format.

VULNERABILITY ASSESSMENT: CHECKLIST PROCEDURE

In performing the vulnerability assessment of any Communications Sector unit or facility, one of the simplest methodologies to employ is the checklist. The Building Vulnerability Assessment Checklist was developed by the Department of Veterans Affairs (VA) and is part of EEMA 426, *Reference Manual to Mitigate Potential Terrorist Attacks Against Buildings*, the reference manual that is recommended in this book and is available on the Internet at http://www.fema.gov/media-library-data/2013072. It is an excellent guide for conducting a viable Checklist-Type Vulnerability Assessment. This Checklist will help you to prepare your Threat Assessment because it allows a consistent security evaluation of designs at various levels. The Checklist can be used as a screening tool for preliminary design vulnerability assessment and supports the preparation of all steps for use by the assessment teams during preparation for interviews with facility representatives to help assure that all relevant aspects of the communications assets are considered in the survey.

REFERENCES AND RECOMMENDED READING

CBO. (2004). *Homeland Security and the Private Sector*. Accessed 05/2/168. www.cbo.gov/ft.docs.

CSI. (2011). *2010/2011 Computer Crime and Security Survey*. Orlando, FL: Computer Security Institute.

DHS. (2009). *National Infrastructure Protection Plan*. Accessed 05/11/19. http://www.dhs.gov/xlibrary/assets/NIPP.Plan.pdf.

DHS. (2010). Communications Sector-Specific Plan. Washington, DC: U.S. Department of Homeland Security.

DoDD. (2010). DoD Policy and Responsibility for Critical Infractions -DODD 3020.40. Washington, DC: U.S. Department of Defense.

FBI. (2012). Insider Espionage. Report before the House Committee on Homeland Security Subcommittee on Counter Terrorism and Intelligence. Washington, DC: Federal Bureau of Investigation.

FEMA. (2008). *FEMA452: Risk Assessment: A How-to Guide*. Accessed 05/01/16. fema.gov/library/file?type=published/filetofile.

FEMA. (2015). *Protecting Critical Infrastructure Against Insider Threats*. Accessed 17 April 2015. http://emilms.fema.gov/IS0915/IABsummary.htm.

INSA. (2017). *Building a Stronger Intelligence Community*. Arlington, VA: Intelligence and National Security Alliance. www.insooline.org.

National Infrastructure Advisory Council. (2008). *First Report and Recommendations on the Insider Threat to Critical Infrastructure*. Washington, DC.

Spellman, F. R. (1997). *A Guide to Compliance for PSM/RMP*. Lancaster, PA: Technomic Publishing Company.

U.S. Department of Energy. 2010. *Energy Sector-Specific Plan: An Annex to the National Infrastructure Protection Plan*. Washington, DC: USDOE.

US DOE. (2002). *Vulnerability Assessment Methodology: Electric Power Infrastructure*. Washington, DC.

US EPA. 2002. *Vulnerability Assessment Fact Sheet*. EPA 816-F-02-025. Accessed May 2006. www.epa.gov/ogwdw/security/index.html.

6

Preparation

When Is Enough, Enough?

Question: When preparing to respond to terrorist acts against people, malls, schools, sports stadiums, libraries, and government facilities, when is enough, enough?

Answer: Until we can read the terrorists' minds, "enough" preparation is never enough. Simply, preparation is ongoing and never-ending.

—Frank R. Spellman

The Headline (09/18/16)

NEW YORK BOMBING: INVESTIGATORS SEARCH FOR SUSPECTS, MOTIVE

—CNN headline, 2016

We must take the battle to the enemy, disrupt his plans, and confront the worst threats before they emerge.

—George W. Bush

Governor Tom Ridge points out the security role for public service professionals:

> Americans should find comfort in knowing that millions of their fellow citizens are working every day to ensure our security at every level—federal, state, county, municipal. These are dedicated professionals who are good at what they do. I've seen it up close, as Governor of Pennsylvania. . . . But there may be gaps in the system. The job of the Office of Homeland Security will be to identify those gaps and work to close them.

> Now, obviously, the further removed we get from September 11, I think the natural tendency is to let down our guard. Unfortunately, we cannot do that. The government will continue to do everything we can to find and stop those who seek to harm us. And I believe we owe it to the American people to remind them that they must be vigilant, as well. (Henry, 2002)

INTRODUCTION

The possibility of terrorism—attacks on U.S. Communications Sector infrastructure—doesn't generate the same attention as potential nuclear, biological or chemical terrorism. Why? Simply, when you ask a citizen about U.S. Communications Sector assets, he or she just looks at you like you are some kind of wacko. I know. I have asked. However, because of the seriousness of the threat of terrorism to the nation's communications sector and the enormous economic and security implications of such attacks, US FCC, USDOE, USEPA, and other agencies have worked non-stop since 9/11 in gathering and providing as much advice and guidance as possible to aid communications sector personnel in protecting communications sector assets and associated critical support infrastructure. In this chapter, we provide an overview of important tools that can be used in protecting communications sector to guard against the threat of terrorism. In the discussion, keep in mind that even though a communications sector issue, the guidance provided could be used to protect the other critical infrastructure sectors.

THREATS AND INCIDENTS

Based on evidence of potential losses from past accidents, indication of the potential human and environmental losses and economic costs from an attack on a communications sector facility or producer comes from major incidents that have occurred both abroad and in the United States. Those events indicate that the human and environmental losses could be significant (CBO, 2004).

Communication sector threats and incidents may be of particular concern due to the range of potential consequences:

- Creating an adverse impact on public health within a population.
- Disrupting system operations and interrupting the supply of critical military components.
- Causing physical damage to system infrastructure.

- Reducing public confidence in the communications supply system.
- Denying of long-term basic security and protection and the cost of replacement.

Keep in mind that some of these consequences would only be realized in the event of a successful terrorist incident; however, the mere threat of terrorism can also have an adverse impact on industries that depend on a safe, steady supply of communication transmission and supplies. In addition, the economic implications of such attacks are potentially enormous. For example, many believe that the reason we are looking at oil at more than $60 a barrel is the fact that we have a "terror premium" factored into the price of a barrel of oil.

While it is important to consider the range of possibilities associated with a particular threat, assessments are typically based on the probability of a particular occurrence. Determining probability is somewhat subjective, and is often based on intelligence and previous incidents. As mentioned, there are historical accounts of accidental incidents that have caused tremendous death and destruction.

Threat Warning Signs

A threat warning is an occurrence or discovery that indicates a potential threat that triggers an evaluation of the threat. It is important to note that these warnings must be evaluated in the context of typical industry activity and previous experience in order to avoid false alarms. Following is a brief description of potential warnings.

- *Security Breach.* Physical security breaches, such as unsecured doors, open hatches, and unlocked/forced gates, are probably the most common threat warnings. In most cases, the security breach is likely related to lax operations or typical criminal activity such as trespassing, vandalism, and theft. However, it may be prudent to assess any security breach with respect to the possibility of attack.
- *Witness Account.* Awareness of an incident may be triggered by a witness account of tampering. Communications sector sites/facilities should be aware that individuals observing suspicious behavior near communications sector facilities/plants will likely call 911 and not the plant. In this

case, the incident warning technically might come from law enforcement, as described below. Note: the witness may be a plant employee engaged in their normal duties.

- *Direct Notification by Perpetrator.* A threat may be made directly to the communications sector site, plant or facility, either verbally or in writing. Historical incidents would indicate that verbal threats made over the phone are more likely than written threats. While the notification may be a hoax, threatening a communications sector unit is a crime and should be taken seriously.
- *Notification by Law Enforcement.* A communications sector site/facility may receive notification about a threat directly from law enforcement, including local, county, state, or federal agencies. As discussed previously, such a threat could be a result of suspicious activity reported to law enforcement, either by a perpetrator, a witness, or the news media. Other information, gathered through intelligence or informants, could also lead law enforcement to conclude that there may be a threat to the communication sector site/facility. While law enforcement will have to lead the criminal investigation, the communication sector site/facility has primary responsibility for the safety of it equipment, energy-producing materials and processes and public health. Thus, the plant's role will likely be to help law enforcement to appreciate the public health implications of a particular threat as well as the technical feasibility of carrying out a particular threat.
- *Notification by News Media.* A threat to destroy a communication sector site/facility might be delivered to the news media, or the media may discover a threat. A conscientious reporter would immediately report such a threat to the police, and either the reporter or the police would immediately contact the communication sector site/facility. This level of professionalism would provide an opportunity for the plant to work with the media and law enforcement to assess the credibility of the threat before any broader notification is made.

RESPONSE TO THREATS

Note: This section is not designed to discuss what specific steps to take in responding to a terrorist threat. Rather, the questions addressed in this section are "Why is it necessary to plan to respond to communications sector threats at all?" and "When have I done enough?"

Federal, state, and local programs already exist that—with varying degrees of effectiveness—encourage or require the operators of communications sector sites/facilities to boost their efforts to promote safety and security and to share information that can help local governments plan for emergencies.

Proper planning is a delicate process because public health measures are rarely noticed or appreciated (like buried utility pipes, they are often hidden functions) except when they fail—then they are very visible. Defense Department consumers are particularly upset by unreliable communications systems and communications services—or practices that produce unsafe (contaminated) environmental media—water, air and/or soil—because they are often viewed as entitlements, and indeed, it is reasonable for citizens to expect a high-quality, safe environment. Public health failures during a response to contamination threats often take the form of too much or too little action. The results to too little action, including no response at all, can have disastrous consequences, potentially resulting in public injuries or fatalities. Considering the potential risks of an inappropriate response to a severe communications sector failure, it is clear that a systematic approach is needed to evaluate contamination threats. One overriding question is "When has a communications producer done enough?" This question may be particularly difficult to address when considering the wide range of agencies that may be involved in a threat situation. Other organizations, such as US FCC, USEPA, USDOE, CDC, USDOT, law enforcement agencies, public health departments, and so forth, will each have unique obligations or interests in responding to a severe release or explosion threat.

When Is Enough, Enough?

The guiding principle for responding to severe release or explosion threats is one of "due diligence" or "a suitable and sensible response to a threat?" As discussed above, some response to communications sector failures is warranted due to the public health implications of an actual dangerous incident.

Ultimately, the answer to the question of "due diligence" must be decided at the local level and will depend on a number of considerations. Among other factors, local authorities must decide what level of risk is reasonable in the context of a perceived threat. Careful planning is essential to developing an appropriate response to terrorist threats, and in fact, one primary objective of USEPA's *Response Protocol Tool Boxes* (RPTBs) is to aid users in the development of their own site-specific plans that are consistent with the needs and responsibilities of the user. Beyond planning, the RPTB considers a careful evaluation of any terrorist threat, and an appropriate response based on the evaluation, to be the most important element of due diligence.

In the RPTB, the threat management process is considered in three successive stages: "possible," "credible," and "confirmed." Thus, as the threat escalates through these three stages, the actions that might be considered due diligence expand accordingly. The following paragraphs describe, in general terms, actions that might be considered as due diligence at these various stages.

- Stage 1: "Is the threat possible?" If a communications facility is faced with a terrorism threat, it should evaluate the available information to determine whether or not the threat is "possible" (i.e., could something have actually happened). If the threat is "possible," immediate operational response actions might be implemented, and activities such as site characterization would be initiated to collect additional information to support the next stage of the threat evaluation.
- Stage 2: "Is the threat credible?" Once a threat is considered "possible," additional information will be necessary to determine if the threat is "credible." The threshold at the credible stage is higher than that at the possible stage, and in general there must be information to corroborate the threat in order for it to be considered "credible."
- Stage 3: "Has the incident been confirmed?" Confirmation implies that definitive evidence and information have been collected to establish the presence of a threat to the communications sector. Obviously, at this stage the concept of due diligence takes on a whole new meaning since authorities are now faced with death and destruction and a potential public health crisis. Response actions at this point include all steps necessary to protect public health, property and the environment.

PREPARATION

As an environmental/occupational safety, security and health professional consultant for various utilities and others on the East Coast, I have performed numerous security, pre-OSHA audit inspections and audits of various plant or facility PSM/RMP compliance programs. During these site visits, one factor always seemed to be universal. While conducting the plant/facility walk-around to gauge the organization's overall profile and status with FCC, OSHA and USEPA compliance, I almost always find that the plant/facility managers or superintendents who accompanied us were shocked to find out what was actually going on or taking place in their facilities. They would scratch their heads and ask various workers: "What the hell are you doing? Where did that new computer come from? When was it installed? And also, why is that door broken? Who told you to paint that door, machine or other apparatus? When did that hole get in the fence? Who left the back gate open? Where is the foreman," and so forth. Eventually, in an expression of utter consternation, the manager/superintendent asks, "Who the hell is in charge around here?" And this is basically the question I find myself asking—far too often.

In one inspection performed at a plant/facility right after 9/11, I drove up to the entrance gate and was impressed with the height and condition of the barbed-razor-wire-topped fence and gates. I could not enter through the gates until I identified myself over a speaker system while a CCTV camera focused on my face. I was let in and given instructions to sign in at the main office. Not bad, just the way it should be . . . or so I thought at the time. After walking most of the plant site in the company of the plant manager, we approached the back fence area which was close to a huge chlorine storage building. At the terminal end of the plant and fence, I noticed a large gate that was propped open with ivy growing on it, through it, and around it—obviously, the gate had been in the open position for quite some time. I asked the plant manager why the gate was open. He stated that it was always open . . . that it led to a downhill path to a beach area below where plant personnel had constructed a picnic area, fronting the James River.

We walked the path to the bottom picnic area, looked around, and then looked back up the path toward the open gate and the prominent structure standing within, the chlorine storage building. While walking back up the path to the gate, I asked the plant manager if he was not concerned about the security and safety of the plant, because the gate was left open, and especially about the safety of the fifty tons of deadly chlorine gas stored in the chlorine storage building.

"Nah . . . no way . . . we are safe here. I really don't see anyone swimming upriver just to get into the plant site, ha. Besides, we are surrounded by woods out here . . . there's nothing to attack anyway."

Once inside the plant, I asked the plant manager if he was not worried about terrorists or some disgruntled former employee using a boat filled with explosives or some other weapon(s) gaining easy access to the plant and especially the chlorine building via the James River beach landing and picnic area.

"Nah, that will never happen . . . who would be that stupid? There's nothing around here worth blowin' up!"

Later, when I checked the GIS/GPS system data and maps pertaining to and showing the plant and surrounding area, I noted that about one-half mile from the plant site was a large housing area, a brewery (with 700+ employees), and a very large theme and historical park—annually visited by more than 2,500,000 people each summer.

Know Your Communication Sector Production Systems

All communications sector facility managers and communications equipment operating personnel must know their plants/facilities. For these persons, there is no excuse for not knowing every square inch of the plant/facility site. In particular, plant workers should know about any and all construction activities underway on the site; the actual construction parameters of the plant; and especially the operation of all plant unit processes. In addition, plant management must not only know their operating staff but also their customers.

Construction and Operation

Each communications sector facility is unique with respect to age, operation, and complexity. Another aspect to be considered relates to the material-producing/using and manufacturing processing system. This is important, particularly in evaluating the potential of a communications failure or malicious action causing and spreading a spilled hazardous material or other potential contaminant. Propagation of production-chemical contaminants through a system is dependent on a number of factors, including mixing conditions at the point of contamination, hydraulic conditions within water systems at the time of the contaminant introduction, reactions between the contaminant and other materials in the system or environment, and electronic control systems operated via a communication system or network.

Information about construction materials used in the system may be contained within the plant records and can be useful in evaluating the fate and transport of a particular chemical contaminant through a system. For example, a particular contaminant may adsorb to the pipe material used in a utility's distribution system, and this type of information would be critical in evaluating remediation options following a chemical contamination incident.

Personnel

Communication sector employees are generally its most valuable asset in preparing for and responding to threats and incidents. They have knowledge of the system and potential problem areas. The importance of knowledgeable and experienced personnel is highlighted by the complexity of most communications systems. This complexity makes a specific terrorist target contingent upon detailed knowledge of the system configuration and usage patterns. If perpetrators have somehow gained a sophisticated understanding of a communications network system, the day-to-day experience of network production will prove an invaluable tool to countering any attacks. For instance, personnel may continually look for unusual aspects of daily operation that might be interpreted as a potential threat warning, and may also be aware of specific characteristics of the system that make it vulnerable to malware attacks or worse.

Customers

A customer's knowledge of communication availability, functionality and delivery is an important component of preventing and managing system intrusion incidents. Prevention is based largely on understanding potential types of malware attacks and the type of target facility. Of special concern may be government buildings and military installations, or other institutions where large numbers of people could be directly or indirectly affected by a communications malfunction threat or incident. Steps taken to protect the customer's communications tie-ins and in addition its employees and property, such as enhancements to the physical security of the sender's and receiver's communications system may deter the attack itself.

Communication customers vary significantly with regard to their expectations of what constitutes acceptable service, so it is necessary to consider the manner in which communications sector services are used in a particular system. For example, some customers, such as factories and product manufacturers, may have certain cyber control system operations to control chemical quality requirements. That being said, for the purposes of managing manufacturing, chemical, fire, explosion, and contamination threats, it is important to keep in mind that the most important goal is protecting the health of the public as a whole. Planning, preparation, and allocation of communication resources should be directed toward protecting the public at large, beyond specific demographic groups or individual users.

Perform Training and Desk/Field Exercises

In addition to a lack of planning, another reason that emergency response plans fail is lack of training and practice. Training provides the necessary means for everyone involved to acquire the skills to fulfill their role during an emergency. It may also provide important "buy-in" to the response process from both management and staff, which is essential to the success of any response plan. Desk exercises (also known as "tabletops" or "sand lot" or "dry runs") along with field exercises allow participants to practice their skills. Also, these exercises will provide a test of the communication security plan itself, revealing strengths and weakness that may be used to improve the overall plan.

Enhance Physical Security

Where possible deny physical access to key sites; within the communications sector system this may act as a deterrent to a perpetrator. When we consider that many of the communication sector units such as power generating and tower transmission systems are often in remote, wide-open spaces, this can be a huge challenge. Terrorists often seek the easiest route of attack, just like a burglar prefers a house with an open window or an open automobile with keys in the ignition. Aside from deterring actual attacks, enhancing physical security has other benefits. For example, installation of fences and locks may reduce the rate of false alarms. Without surveillance equipment or locks, it may not be possible to determine whether a suspicious individual has actually entered a vulnerable area. The presence of a lock and a determination as to whether it has been cut or broken provides sound, although not definitive, evidence that an intrusion has occurred. Likewise, security cameras can be used to review security breaches and determine if the incident was simply due to trespassing or is a potential contamination threat. The costs of enhancing physical security may be justified by comparison to the cost of responding to just one "credible" munitions explosion or contamination threat involving site characterization and lab analysis for potential contaminants.

THE BOTTOM LINE

This chapter has emphasized the importance of ensuring the physical security of communications sector facilities and equipment. But it is important to point out that true communications sector security goes beyond the physical plant; the sector requires security of a different kind. Much of the communications infrastructure, including control architecture, is vulnerable to cyber attack from either inside or outside the network. The fact is that perhaps the largest vulnerability and dependency of the national communications infrastructure as well as the infrastructure for other sectors is on cybersecurity. The control of communications networks and all of their functional components are vulnerable to various degrees of cyber attacks on the software operating systems by either the idle hacker of the more malicious intruder participating in information warfare. This is also true of all the networks within others sectors. The bottom line, the networks must be protected and guarded from attack.

REFERENCES AND RECOMMENDED READING

Henry, K. (2002). New face of security. *Government Security*, pp. 30–37.

U.S. Department of Energy. (2010). *Energy Sector-Specific Plan: An Annex to the National Infrastructure Protection Plan.* Washington, DC: USDOE.

USEPA. (2001). *Protecting the Nation's Water Supplies from Terrorist Attack: Frequently Asked Questions.* Washington, DC: United States Environmental Protection Agency.

USEPA. (2004). *Response Protocol Toolbox: Planning for and Responding to Drinking Water Contamination Threats and Incidents.* Washington, DC: United States Environmental Protection Agency.

USEPA. (2011). *Response Protocol Toolbox: Planning for and Responding to Wastewater Contamination Threats and Incidents.* Washington, DC: United States Environmental Protection Agency.

7

Cybersecurity and SCADA

What will you do when you are hacked?
What will you do when the power goes out?

Nobody has ever been killed by a cyberterrorist.

Unless people are injured, there is also less drama and emotional appeal.

—Dorothy Denning

On April 23, 2000, police in Queensland, Australia stopped a car on the road and found a stolen computer and radio inside. Using commercially available technology, a disgruntled former employee had turned his vehicle into a pirate command center of sewage treatment along Australia's Sunshine Coast. The former employee's arrest solved a mystery that had troubled the Maroochy Shire wastewater system for two months. Somehow the system was leaking hundreds of thousands of gallons of putrid sewage into parks, rivers and the manicured grounds of a Hyatt Regency hotel—marine life died, the creek water turned black and the stench was unbearable for residents. Until the former employee's capture—during his 46th successful intrusion—the utility's managers did not know why.

Specialists study this case of cyber-terrorism because, at the time, it was the only one known in which someone used a digital control system deliberately to cause harm. The former employee's intrusion shows how easy it is to break in—and how restrained he was with his power.

To sabotage the system, the former employee set the software on his laptop to identify itself as a pumping station, and then suppressed all alarms. The former employee was the "central control station" during his intrusions, with unlimited command of 300 SCADA nodes governing sewage and drinking water alike.

The bottom line: as serious as the former employee's intrusions were, they pale in comparison with what he could have done to the fresh water system—he could have done anything he liked.

—Barton Gellman, 2002

Other reports of cyber exploits illustrate the debilitating effects such attacks can have on the nation's security, economy, and public health and safety.

- In May 2015, media sources reported that data belonging to 1.1 million health insurance customers in the Washington, D.C. area were stolen in a cyber attack on a private insurance company. Attackers accessed a database containing names, birth dates, e-mail addresses, and subscriber ID numbers of customers.
- In December 2014, the Industrial Control Systems Cyber Emergency Response Team (ICS-CERT, works to reduce risks within and across all critical infrastructure sectors by partnering with law enforce agencies) issued an updated alert on a sophisticated malware campaign compromising numerous industrial control system environments. Their analysis indicated that this campaign had been ongoing since at least 2011.
- In the January 2014 to April 2014 release of its Monitor report, ICS-CERT reported that a public utility had been compromised when a sophisticated threat actor gained unauthorized access to its control system network through a vulnerable remote access capability configured on the system. The incident highlighted the need to evaluate security controls employed at the perimeter and ensure that potential intrusion vectors are configured with appropriate security controls and monitoring and detection capabilities.

In 2000, the FBI identified and listed threats to critical infrastructure. These threats are listed and described in Table 7.1. In 2015, the GAO described the sources of cyber-based threats. These threats are listed and described in detail in Table 7.2.

IN THE WORDS OF MASTER SUN TZU FROM "THE ART OF WAR":

Those who are first on the battlefield, and await the opponents are at ease; those who are last, and head into battle are worn out.

Table 7.1. Threats to Critical Infrastructure Observed by the FBI

Threat	*Description*
Criminal groups	There is an increased use of cyber intrusions by criminal groups who attack systems for purposes of monetary gain.
Foreign intelligence services	Foreign intelligence services use cyber tools as part of their information-gathering and espionage activities.
Hackers	Hackers sometimes crack into networks for the thrill of the challenge or for bragging rights in the hacker community. While remote cracking once required a fair amount of skill or computer knowledge, hackers can now download attack scripts and protocols from the Internet and launch them against victim sites. Thus, while attack tools have become more sophisticated, they have also become easier to use.
Hacktivists	Hacktivism refers to politically motivated attacks on publicly accessible Web pages or e-mail servers. These groups and individuals overload e-mail servers and hack into websites to send a political message.
Information warfare	Several nations are aggressively working to develop information warfare doctrine, programs, and capabilities. Such capabilities enable a single entity to have a significant and serious impact by disrupting the supply, communications, and economic infrastructures that support military power—impacts that, according to the Director of Central Intelligence, can affect the daily lives of Americans across the country.
Inside threat	The disgruntled organization insider is a principal source of computer crimes. Insiders may not need a great deal of knowledge about computer intrusions because their knowledge of a victim system often allows them to gain unrestricted access to cause damage to the system or to steal system data. The insider threat also includes outsourcing vendors.
Virus writers	Virus writers are posing an increasingly serious threat. Several destructive computer viruses and "worms" have harmed files and hard drives, including the Melissa Macro Virus, Explore, Zip worm, CIH (Chernobyl) Virus, Nimda, and Code Red.

Source: FBI (2000, 2014).

DID YOU KNOW?

Presidential Policy Directive 21 defined "all hazards" as a threat to an incident natural or manmade that warrants action to protect life, property, the environment, and public health or safety, and to minimize disruptions of government, social, or economic activities.

Threats to systems supporting critical infrastructure are evolving and growing. As shown in Table 7.2, cyber threats can be unintentional or intentional. Unintentional or non-adversarial threats include equipment failures, software coding errors, and the actions of poorly trained employees. They also include natural disasters and failures of critical infrastructure on which the organization depends but are outside of its control. Intentional threats include both targeted and untargeted attacks form a variety of sources, including criminal groups, hackers, disgruntled employees, foreign nations engaged in espionage and information warfare, and terrorists. These threat adversaries vary in terms of the capabilities of the actors, their willingness to act, and their motives, which can include seeking monetary gain or seeking an economic, political or military advantage (GAO, 2015).

CYBERSPACE

Today's developing "information age" technology has intensified the importance of critical infrastructure protection, in which cybersecurity has become as critical as physical security to protecting virtually all critical infrastructure sectors. The Department of Defense (DoD) determined that cyber threats to contractors' unclassified information systems represented an unacceptable risk of compromise to DoD information and posed a significant risk to U.S. national security and economic security interests.

In the past few years, especially since 9/11, it has been somewhat routine for us to pick up a newspaper or magazine or view a television news program where a major topic of discussion is cybersecurity or the lack thereof. For

Table 7.2. Common Cyber Threat Sources

Source	*Description*
Non-adversarial-malicious	
Failure in information technology equipment	Failures in displays, sensors, controllers, and information technology hardware responsible for data storage, processing, and communications
Failure in environmental controls	Failures in temperature/humidity controllers or power supplies
Software coding errors	Failures in operating systems, networking, and general-purpose and mission-specific applications
Natural or man-made disaster	Events beyond an entity's control such as fires, floods/tsunamis, tornadoes, hurricanes, and earthquakes
Unusual or natural event	Natural events beyond the entity's control that are not considered to be disasters (e.g., sunspots)
Infrastructure failure or outage	Failure or outage of telecommunications or electrical power
Unintentional user errors	Failures resulting from erroneous, accidental actions taken by individuals (both system users and administrators) in the course of executing their everyday responsibilities
Adversarial	
Hackers or hacktivists	Hackers break networks for the challenge, revenge, stalking, or monetary gain, among other reasons. Hactivists are ideologically motivated actors who use cyber exploits to further political goals.
Malicious insiders	Insiders (e.g., disgruntled organization employees, including contractors) may not need a great deal of knowledge about computer intrusions because their position with the organization often allows them to gain unrestricted access and cause damage to the target system or to steal system data. These individuals engage in purely malicious activities and should not be confused with non-malicious insider accidents.
Nations	Nations, including nation-state, state-sponsored, and state sanctioned programs use cyber tools as part of their information-gathering and espionage activities. In addition, several nations are aggressively working to develop information warfare doctrine, programs, and capabilities.
Criminal groups and organized crime	Criminal groups seek to attack systems for monetary gain. Specifically, organized criminal groups use cyber exploits to commit identity theft, online fraud, and computer extortion.
Terrorists	Terrorists seek to destroy, incapacitate, or exploit critical infrastructures in order to threaten national security, cause mass casualties, weaken the economy, and damage public morale and confidence.
Unknown malicious outsiders	Unknown malicious outsiders are threat sources or agents that, due to a lack of information, agencies are unable to classify as being one of the five types of threat sources or agents listed above.

Source: GAO Analysis of Unclassified Government and Nongovernmental Data. GAO 16-79.

example, recently there has been discussion about Russian hackers trying to influence the U.S. 2016 elections. Many of the cyber intrusion incidents we read or hear about have added new terms or new uses for old terms to our vocabulary. For example, old terms such as Botnets (short for robot networks, also called bots, zombies, botnet fleets, and many others) are groups of computers that have been compromised with malware such as Trojan Horses, worms, backdoors, remote control software, and viruses have taken on new connotations in regard to cybersecurity issues. Relatively new terms such as scanners, Windows NT hacking tools, ICQ hacking tools, mail bombs, sniffer, logic bomb, nukers, dots, backdoor Trojan, key loggers, hackers' Swiss knife, password crackers, blended threats, Warhol Worms, Flash Threats, Targeted Attacks, and BIOS crackers are now commonly read or heard. New terms have evolved along with various control mechanisms. For example, because many control systems are vulnerable to attacks of varying degrees, these attack attempts range from telephone line sweeps (wardialing), to wireless network sniffing (wardriving), to physical network port scanning, and to physical monitoring and intrusion. Wireless network sniffing is performed at (or near) the target point by a pedestrian (warwalking), meaning that instead of a person being in an automotive vehicle, the potential intruder may be sniffing the network for weaknesses or vulnerabilities on foot, posing as a person walking, but they may have a handheld PDA device or laptop computer (Warwalking 2003). Further, adversaries can leverage common computer software programs, such as Adobe acrobat and Microsoft Office, to deliver a threat by embedding exploits within software files that can be activated when a user opens a file within its corresponding program. Finally, the communications infrastructure and the utilities are extremely dependent on the Information Technology Sector. This dependency is due to the reliance of the communications systems on the software that runs the control mechanism of the operations systems, the management software, the billing software, and any number of other software packages used by industry. Table 7.3 provides descriptions of common exploits or techniques, tactics, and practices used by cyber adversaries.

Table 7.3. Common Methods of Cyber Exploits

Exploit	*Description*
Watering hole	A method by which threat actors exploit the vulnerabilities of carefully selected websites frequented by users of the targeted system. Malware is then injected to the targeted system via the compromised websites.
Phishing and spear phishing	A digital form of social engineering that uses authentic-looking e-mails, websites, or instant messages to get users to download malware, open malicious attachments, or open links that direct them to a website that requires information or executes malicious code.
Credentials based	An exploit that takes advantage of a system's insufficient user authentication and/or any elements of cybersecurity supporting it, to include not limiting the number of failed login attempts, the use of hard-coded credentials, and the use of a broken or risky cryptographic algorithm.
Trusted third parties	An exploit that takes advantage of the security vulnerabilities of trusted third parties to gain access to an otherwise secure system.
Classic buffer overflow	An exploit that involves the intentional transmission of more data than a program's input buffer can hold, leading to the deletion of critical data and subsequent execution of malicious code.
Cryptographic weakness	An exploit that takes advantage of a network employing insufficient encryption when either storing or transmitting data, enabling adversaries to read and/or modify the data stream.
Structured Query Language (SQL) injection	An exploit that involves the alteration of a database search in a web-based application, which can be used to obtain unauthorized access to sensitive information in a database, resulting in data loss at corruption, denial or service, or complete host takeover.
Operating system command injection	An exploit that takes advantage of a system's inability to properly neutralize special elements used in operating system commands, allowing the adversaries to execute unexpected commands on the system by either modifying already evoked commands or evoking their own.

(*Continued*)

Table 7.3. (Continued)

Exploit	*Description*
Cross-site scripting	An exploit that uses third-party web resources to run lines of programming code (referred to as scripts) within the victim's web browser or scriptable application. This occurs when a user, using a browser, visits a malicious website or clicks a malicious link. The most dangerous consequences can occur when this method is used to exploit additional vulnerabilities that may permit an adversary to steal cookers (data exchanged between a web server and a browser), log key stokes, capture screen shots, discover and collect network information, or remotely access and control the victim's machine.
Cross-site request forgery	An exploit takes advantage of an application that cannot, or does not, sufficiently verify whether a well-formed, valid, consistent request was intentionally provided by the user who submitted the request, tricking the victim into executing a falsified requirement that results in the system or data being compromised.
Path traversal	An exploit that seeks to gain access to files outside of a restricted directory by modifying the directory pathname in an application that does not properly neutralize special elements (e.g., ' . . . ', '/', ' . . . /', etc.)
Integer overflow	An exploit where malicious code is inserted that leads to unexpected integer overflow, or wraparound, which can be used by adversaries to control looping or make security decisions in order to cause program crashes, memory corruption, or the execution of arbitrary code via buffer overflow.
Uncontrolled format string	Adversaries manipulate externally controlled format strings in print-style functions to gain access to information and/or execute unauthorized code or commands.
Open redirect	An exploit where the victim is tricked into selecting a URL (website location) that has been modified to direct them to an external, malicious site which may contain malware that can compromises the victim's machine.

Exploit	Description
Heap-based buffer overflow	Similar to classic buffer overflow, but the buffer that is overwritten is allocated in the heap portion of memory, generally meaning that the buffer was allocated using a memory allocation routine, such as "malloc ( )."
Unrestricted upload of files	An exploit that takes advantage of insufficient upload restrictions, enabling adversaries to upload malware (e.g., .php) in place of the intended file type (e.g., .jpg).
Inclusion of functionality from untrusted sphere	An exploit that uses trusted, third-party executable functionality (e.g., web widget or library) as a means of executing malicious code in software whose protection mechanisms are unable to determine whether functionality is from trusted sources, modified in transit, or being spoofed.
Certificate and certificate authority compromise	Exploits facilitated via the issuance of fraudulent digital certificates (e.g., transport layer security and Secure Socket Layer). Adversaries use these certificates to establish secure connections with the target organization or individual by mimicking a trusted third party.
Hybrid of others	An exploit which combines elements of two or more of the aforementioned techniques.

Source: GAO (2015).

Not all relatively new and universally recognizable cyber terms have a sinister connotation or meaning, of course. Consider, for example, the following digital terms: backup, binary, bit byte, CD-ROM, CPU, database, e-mail, HTML, icon, memory, cyberspace, modem, monitor, network, RAM, Wi-Fi (wireless fidelity), record, software, World Wide Web—none of these terms normally generate thoughts of terrorism in most of us.

There is, however, one digital term, SCADA, that most people have not heard of. This is not the case, however, for those who work with the nation's critical infrastructure, including communications sector infrastructure. SCADA, or **S**upervisory **C**ontrol **A**nd **D**ata **A**cquisition System (also sometimes referred to as Digital Control Systems or Process Control Systems), provides the real-time control mechanisms for most of the power and utility facilities and plays an important role in computer-based control systems. From coordinating music and lights in the proper sequence with controlling vaulting of spray from water fountains to controlling systems used in the drilling and refining of oil and natural gas, control systems perform many functions. Many energy distribution networks use computer-based systems to remotely control sensitive feeds, processes and system equipment previously controlled manually. These systems (commonly known as SCADA) allow an energy utility (or operation) to monitor fuel tank levels, to ensure that contents are stored at correct levels or to monitor tank levels, and, as mentioned, collect data from sensors and control equipment located at remote sites. Common communication sector and customer system sensors measure many parameters, depending on the nature of the industrial manufacturing entity. Common SCADA customer industry system equipment includes valves, pumps, and switching devices for distribution of electricity. The critical infrastructure of many countries is increasingly dependent on SCADA systems.

WHAT IS SCADA?

If we were to ask the specialist to define SCADA, the technical response could be outlined as follows:

SCADA is:

- A multi-tier system or interfaces with multi-tier systems
- Used for physical measurement and control endpoints via Remote Terminal Unit (RTU) and Program Logic Controller (PLC) to measure voltage, adjust a value or flip a switch
- An intermediate processor normally based on commercial 3rd party OSes—VMS, Unix, Windows, Linux
- Interfaces, for example, with a graphical user interface (Windows GUIs)
- Communication infrastructure that consists of a variety of transport mediums such as analog, serial, Internet, radio, Wi-Fi

How about the non-specialist response?—for the rest of us who are non-specialists. Well, for those of us in this category we could simply say that SCADA is a computer-based control system that remotely controls and monitors processes previously controlled and monitored manually. The philosophy behind SCADA control systems can be summed up by the phrase "If you can measure it, you can control it." SCADA allows an operator using a central computer to supervise (control and monitor) multiple networked computers at remote locations. Each remote computer can control mechanical processes (mixers, pumps, valves, etc.) and collect data from sensors at its remote locations. Thus the phrase: **S**upervisory **C**ontrol **a**nd **D**ata **A**cquisition, or **SCADA**.

The central computer is called the Master Terminal Unit, or MTU. The MTU has two main functions: periodically obtain data from RTUs/PLCs and control remote devices through the operator station. The operator interfaces with the MTU using software called Human Machine Interface, or HMI. The remote computer is called Program Logic Controller (PLC) or Remote Terminal Unit (RTU). The RTU activates a relay (or switch) that turns mechanical equipment "on" and "off." The RTU also collects data from sensors. Sensors perform measurement, and actuators perform control.

In the initial stages utilities ran wires, also known as hardwire or land lines, from the central computer (MTU) to the remote computers (RTUs). Because remote locations can be located hundreds of miles from the central location, utilities began to use public phone lines and modems, leased telephone company lines, and radio and microwave communication. More recently, they have also begun to use satellite links, Internet, and newly developed wireless technologies.

Because SCADA systems' sensors provided valuable information, many critical infrastructure entities, utilities and other industries established "connections" between their SCADA systems and their business system. This allowed utility/industrial management and other staff access to valuable statistics, such

DID YOU KNOW?

Modern RTUs typically use a ladder-logic approach to programming due to its similarity to standard electrical circuits. An RTU that employs this ladder-logic programming is called a Programmable Logic Controller (PLC).

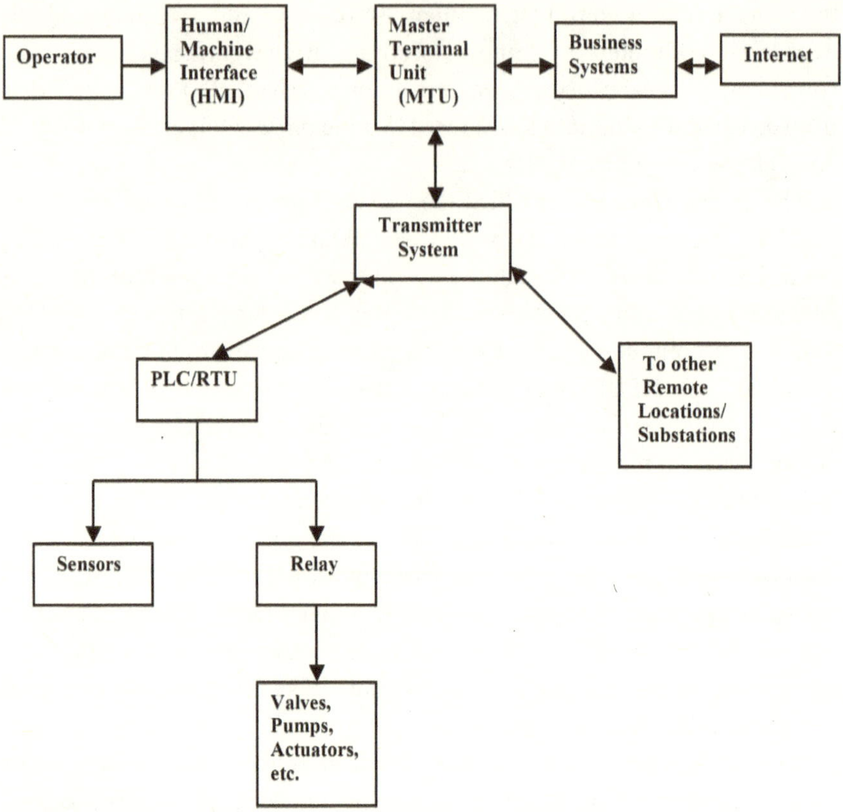

FIGURE 7.1
Representative SCADA network

as chemical usage. When utilities/industries later connected their systems to the Internet, they were able to provide stakeholders/stockholders with usage statistics on the communications segment and utility/industrial web pages. Figure 7.1 provides a basic illustration of a representative SCADA network. Note that firewall protection would normally be placed between the Internet and business system and between the business system and the MTU.

SCADA APPLICATIONS

As stated above, SCADA systems can be designed to measure a variety of equipment operating conditions and parameters or volumes and flow rates or electricity, natural gas and oil or oil and petrochemical mixture quality parameters, and to respond to change in those parameters either by alerting operators or by modifying system operation through a feedback loop system without having personnel physically visit each valve, process or piece of other equipment on a daily basis to check it and/or ensure that it is functioning properly. Automation and integration of large-scale diverse assets required SCADA systems to provide the utmost in flexibility, scalability, openness and reliability. SCADA systems are used to automate certain energy production functions; these can be performed without initiation by an operator. In addition to process equipment, SCADA systems can also integrate specific security alarms and equipment, such as cameras, motion sensors, lights, data from card reading systems, and so forth, thereby providing a clear picture of what is happening at areas throughout a facility. Finally, SCADA systems also provide constant, real-time data on processes, equipment, and location access, so that the necessary response can be made quickly. This can be extremely useful during emergency conditions, such as when energy distribution lines or piping breaks or potentially disruptive chemical reaction spikes appear in chemical processing operations. Currently, it can be said that SCADA has evolved from a simple indicating light and pushbutton control system into a comprehensive operation and handling system for very complex process control and safety shutdown systems. In a nutshell, SCADA results in an oversight system that requires fewer operators.

Today, many common communications sector applications for SCADA systems include, but are not limited to those shown below.

- Boiler controls
- Bearing temperature monitor (electric generators and motors)
- Gas processing
- Plant monitoring
- Plant energy management
- Power distribution monitoring
- Electric power monitoring
- Fuel oil handling system
- Hydroelectric load management
- Petroleum pilot plants
- Plant monitoring
- Process controls
- Process stimulators

- Tank controls
- Utility monitoring
- Safety parameter display systems and shutdown systems
- Tank level control and monitoring
- Turbine controls
- Turbine monitoring
- Virtual annunciator panels
- Alarm systems
- Security equipment
- Event logging

Because these systems can monitor multiple processes, equipment, and infrastructure and then provide quick notification of, or response to, problems or upsets, SCADA systems typically provide the first line of detection for atypical or abnormal conditions. For example, a SCADA system connected to sensors that measure specific machining quality parameters are measured outside of a specific range. A real-time customized operator interface screen could display and control critical systems-monitoring parameters.

The system could transmit warning signals back to the operators, such as by initiating a call to a personal pager. This might allow the operators to initiate actions to prevent power outages or contamination and disruption of the energy supply. Further automation of the system could ensure that the system initiated measures to rectify the problem. Preprogrammed control functions (e.g., shutting a valve, controlling flow, throwing a switch, or adding chemicals) can be triggered and operated based on SCADA utility.

SCADA VULNERABILITIES

U.S. Electric Grid Gets Hacked Into

> The Associated Press (AP) reported April 9, 2009 that spies hacked into the U.S. energy grid and left behind computer programs (Trojan Horses) that would enable them to disrupt service, exposing potentially catastrophic vulnerabilities in key pieces of national infrastructure.

Even though terrorists, domestic and/or foreign, tend to aim their main focus around the critical devices that control actual communications system delivery activities, according to USEPA (2005), SCADA networks were developed with little attention paid to security, making the security of these systems often weak. Studies have found that, while technological advancements introduced vulnerabilities, many communications sector facilities and utilities have spent little time securing their SCADA networks. As a result, many SCADA networks may be susceptible to attacks and misuse. SCADA systems languished in obscurity and this was the essence of their security—that is, until technological developments transformed SCADA from a backroom operation to a front-and-center visible control system.

Remote monitoring and supervisory control of processes began to develop in the early 1960s, and adopted many technological advancements. The advent of minicomputers made it possible to automate a vast number of once manually operated switches. Advancements in radio technology reduced the communication costs associated with installing and maintaining buried cable in remote areas. SCADA systems continued to adopt new communication methods including satellite and cellular. As the price of computers and communications dropped, it became economically feasible to distribute operations and to expand SCADA networks to include even smaller facilities.

Advances in information technology and the necessity of improved efficiency have resulted in increasingly automated and interlinked infrastructures, and created new vulnerabilities due to equipment failure, human error, weather and other natural causes, and physical and cyber attacks. Some areas and examples of possible SCADA vulnerabilities include these (Wiles, et al., 2007):

- Human—People can be tricked or corrupted, and may commit errors.
- Communications—Messages can be fabricated, intercepted, changed, deleted, or blocked.
- Hardware—Security features are not easily adapted to small self-contained units with limited power supplies.
- Physical—Intruders can break into a facility to steal or damage SCADA equipment.
- Natural—Tornadoes, floods, earthquakes, and other natural disasters can damage equipment and connections.
- Software—Programs can be poorly written.

Specific SCADA weaknesses and potential attack vectors include these:

- Does not require any authorization.
- Does not use encryption.
- Does not properly recognize and properly handle errors and exceptions.
- No authentication is required.
- Data could be intercepted.
- Manipulation of data.
- Service denial.
- IP address spoofing—Internet protocol packets with a false source IP address.
- Session hijacking.
- Unsolicited responses.
- Packet fuzzing—involves inputting false information, etc.
- Unauthorized control.
- Log data medication.

SCADA system computers and their connections are susceptible to different types of information system attacks and misuse such as those mentioned above. The Computer Security Institute and Federal Bureau of Investigation conduct an annual Computer Crime and Security Survey (FBI, 2007). The survey reported on ten types of attacks or misuse, and reported that viruses and denial of service had the greatest negative economic impact. The same study also found that 15 percent of the respondents reported abuse of wireless networks, which can be a SCADA component. On average, respondents from all sectors did not believe that their organization invested enough in security awareness. For example, utilities as a group reported a lower average computer security expenditure/investment per employee than many other sectors such as transportation, telecommunications, and financial.

Sandia National Laboratories' *Common Vulnerabilities in Critical Infrastructure Control Systems* described some of the common problems it has identified in the following five categories (Stamp et al., 2003):

1. **System Data**—Important data attributes for security include availability, authenticity, integrity, and confidentiality. Data should be categorized according to its sensitivity, and ownership and responsibility must be assigned. However, SCADA data is often not classified at all, making

it difficult to identify where security precautions are appropriate (for example, which communication links to secure, databases requiring protection, etc.).

2. **Security Administration**—Vulnerabilities emerge because many systems lack a properly structured security policy (security administration is notoriously lax in the case of control systems), equipment and system implementation guides, configuration management, training, and enforcement and compliance auditing.
3. **Architecture**—Many common practices negatively affect SCADA security. For example, while it is convenient to use SCADA capabilities for other purposes such as fire and security systems, these practices create single points of failure. Also, the connection of SCADA networks to other automation systems and business networks introduces multiple entry points for potential adversaries.
4. **Network** (including communication links)—Legacy systems' hardware and software have very limited security capabilities, and the vulnerabilities of contemporary systems (based on modern information technology) are publicized. Wireless and shared links are susceptible to eavesdropping and data manipulation.
5. **Platforms**—Many platform vulnerabilities exist, including default configurations retained, poor password practices, shared accounts, inadequate protection for hardware, and nonexistent security monitoring controls. In most cases, important security patches are not installed, often due to concern about negatively impacting system operation; in some cases technicians are contractually forbidden from updating systems by their vendor agreements.

The following incident helps to illustrate some of the risks associated with SCADA vulnerabilities.

- During the course of conducting a vulnerability assessment, a contractor stated that personnel from his company penetrated the information system of a utility within minutes. Contractor personnel drove to a remote substation and noticed a wireless network antenna. Without leaving their vehicle, they plugged in their wireless radios and connected to the network within 5 minutes. Within 20 minutes they had mapped the network, including

SCADA equipment, and accessed the business network and data. This illustrates what a cybersecurity advisor from Sandia National Laboratories specializing in SCADA stated, that utilities are moving to wireless communication without understanding the added risks.

The Increasing Risk

According to GAO (2015), historically, security concerns about control systems (SCADA included) were related primarily to protecting against physical attack and misuse of refining and processing sites or distribution and holding facilities. However, more recently there has been a growing recognition that control systems are now vulnerable to cyber attacks from numerous sources, including hostile governments, terrorist groups, disgruntled employees, and other malicious intruders.

In addition to control system vulnerabilities mentioned earlier, several factors have contributed to the escalation of risk to control systems, including (1) the adoption of standardized technologies with known vulnerabilities, (2) the connectivity of control systems to other networks, (3) constraints on the implementation of existing security technologies and practices, (4) insecure remote connections, and (5) the widespread availability of technical information about control systems.

Adoption of Technologies with Known Vulnerabilities

When a technology is not well known, not widely used, not understood or publicized, it is difficult to penetrate it and thus disable it. Historically, proprietary hardware, software, and network protocols made it difficult to understand how control systems operated—and therefore how to hack into them. Today, however, to reduce costs and improve performance, organizations have been transitioning from proprietary systems to less expensive, standardized technologies such as Microsoft's Windows and Unix-like operating systems and the common networking protocols used by the Internet. These widely used standardized technologies have commonly known vulnerabilities, and sophisticated and effective exploitation tools are widely available and relatively easy to use. As a consequence, both the number of people with the knowledge to wage attacks and the number of systems subject to attack have increased. Also, common communication protocols and the emerging use of Extensible Markup Language (commonly referred to as XML) can make it easier for a hacker to interpret the content of communications among the components of a control system.

Control systems are often connected to other networks—enterprises often integrate their control system with their enterprise networks. This increased connectivity has significant advantages, including providing decision makers with access to real-time information and allowing engineers to monitor and control the process control system from different points on the enterprise network. In addition, the enterprise networks are often connected to the networks of strategic partners and to the Internet. Further, control systems are increasingly using wide area networks and the Internet to transmit data to their remote or local stations and individual devices. This convergence of control networks with public and enterprise networks potentially exposes the control systems to additional security vulnerabilities. Unless appropriate security controls are deployed in the enterprise network and the control system network, breaches in enterprise security can affect the operation of a control system.

According to industry experts, the use of existing security technologies, as well as strong user authentication and patch management practices, are generally not implemented in control systems because control systems operate in real time, typically are not designed with cybersecurity in mind, and usually have limited processing capabilities.

Existing security technologies such as authorization, authentication, encryption, intrusion detection, and filtering of network traffic and communications require more bandwidth, processing power, and memory than control system components typically have. Because controller stations are generally designed to do specific tasks, they use low-cost, resource-constrained microprocessors. In fact, some devices in the electrical industry still use the Intel 8088 processor, introduced in 1978. Consequently, it is difficult to install existing security technologies without seriously degrading the performance of the control system.

Further, complex passwords and other strong password practices are not always used to prevent unauthorized access to control systems, in part because this could hinder a rapid response to safety procedures during an emergency. As a result, according to experts, weak passwords that are easy to guess, shared, and infrequently changed are reportedly common in control systems, including the use of default passwords or even no password at all.

In addition, although modern control systems are based on standard operating systems, they are typically customized to support control system applications. Consequently, vendor-provided software patches are generally either incompatible or cannot be implemented without compromising service shutting down "always-on" systems or affecting interdependent operations.

Potential vulnerabilities in control systems are exacerbated by insecure connections. Organizations often leave access links—such as dial-up modems to equipment and control information—open for remote diagnostics, maintenance, and examination of system status. Such links may not be protected with authentication of encryption, which increases the risk that hackers could us these insecure connections to break into remotely controlled systems. Also, control systems often use wireless communications systems, which are especially vulnerable to attack, or leased lines that pass through commercial telecommunications facilities. Without encryption to protect data as it flows through these insecure connections or authentication mechanisms to limit access, there is limited protection for the integrity of the information being transmitted.

Public information about infrastructures and control systems is available to potential hackers and intruders. The availability of this infrastructure and vulnerability data was demonstrated by a university graduate student, whose dissertation reportedly mapped every business and industrial sector in the American economy to the fiber-optic network that connects them—using material that was available publicly on the Internet, none of which was classified.

Cyber Threats to Control Systems

There is a general consensus—and increasing concern—among government officials and experts on control systems about potential cyber threats to the control systems that govern our critical infrastructures. As components of control systems increasingly make critical decisions that were once made by humans, the potential effect of a cyber threat becomes more devastating. Such cyber threats could come from numerous sources, ranging from hostile governments and terrorist groups to disgruntled employees and other malicious intruders.

In July 2002, National Infrastructure Protection Center (NIPC) reported that the potential for compound cyber and physical attacks, referred to as "swarming attacks," is an emerging threat to the U.S. critical infrastructure. As NIPC reports, the effects of a swarming attack include slowing or complicating the response to a physical attack. For instance, a cyber attack that disabled the water supply or the electrical system in conjunction with a physical attack could deny emergency services the necessary resources to manage the consequences—such as controlling fires, coordinating actions, and generating light.

Control systems, such as SCADA, can be vulnerable to cyber attacks. Entities or individuals with malicious intent might take one or more of the following actions to successfully attack control systems:

- disrupt the operation of control systems by delaying or blocking the flow of information through control networks, thereby denying availability of the networks to control system operations;
- make unauthorized changes to programmed instructions in PLCs, RTUs, or DCS controllers, change alarm thresholds, or issue unauthorized commands to control equipment, which could potentially result in damage to equipment (if tolerances are exceeded), premature shutdown of processes (such as prematurely shutting down transmission lines), or even disabling of control equipment;
- send false information to control system operators either to disguise unauthorized changes or to initiate inappropriate actions by system operators;
- modify the control system software, producing unpredictable results; and
- interfere with the operation of safety systems.

In addition, in control systems that cover a wide geographic area, the remote sites are often unstaffed and may not be physically monitored. If such remote systems are physically breached, the attackers could establish a cyber connection to the control network.

Securing Control Systems

Several challenges must be addressed to effectively secure control systems against cyber threats. These challenges include (1) the limitations of current security technologies in securing control systems, (2) the perception that securing control systems may not economically justifiable, and (3) the conflicting priorities within organizations regarding the security of control systems.

A significant challenge in effectively securing control systems is the lack of specialized security technologies for these systems. The computing resources in control systems that are needed to perform security functions tend to be quite limited, making it very difficult to use security technologies within control system networks without severely hindering performance.

Securing control systems may not be perceived as economically justifiable. Experts and industry representatives have indicated that organizations may be reluctant to spend more money to secure control systems. Hardening the security of control systems would require industries to expend more resources, including acquiring more personnel, providing training for personnel, and potentially prematurely replacing current systems that typically have a lifespan of about twenty years.

Finally, several experts and industry representatives indicated that the responsibility for securing control systems typically includes two separate groups: IT security personnel and control system engineers and operators. IT security personnel tend to focus on securing enterprise systems, while control system engineers and operators tend to be more concerned with the reliable performance of their control systems. Further, they indicate that, as a result, those two groups do not always fully understand each other's requirements and collaborate to implement secure control systems.

STEPS TO IMPROVE SCADA SECURITY

The President's Critical Infrastructure Protection Board and the Department of Energy (DOE) have developed the steps outlined below to help organizations improve the security of their SCADA networks. DOE (2001) points out that these steps are not meant to be prescriptive or all-inclusive. However, they do address essential actions to be taken to improve the protection of SCADA networks. The steps are divided into two categories: specific actions to improve implementation, and actions to establish essential underlying management processes and policies.

21 Steps to Increase SCADA Security (DOE, 2001)

The following steps focus on specific actions to be taken to increase the security of SCADA networks:

1. **Identify all connections to SCADA networks.**
   Conduct a thorough risk analysis to assess the risk and necessity of each connection to the SCADA network. Develop a comprehensive under-

standing of all connections to the SCADA network, and how well those connections are protected. Identify and evaluate the following types of connections:

- Internal local area and wide area networks, including business networks
- The Internet
- Wireless network devices, including satellite uplinks
- Modem or dial-up connections
- Connections to business partners, vendors or regulatory agencies

2. **Disconnect unnecessary connections to the SCADA network.**
To ensure the highest degree of security of SCADA systems, isolate the SCADA network from other network connections to as great a degree as possible. Any connection to another network introduces security risks, particularly if the connection creates a pathway from or to the Internet. Although direct connections with other networks may allow important information to be passed efficiently and conveniently, insecure connections are simply not worth the risk; isolation of the SCADA network must be a primary goal to provide needed protection. Strategies such as utilization of "demilitarized zones" (DMZs) and data warehousing can facilitate the secure transfer of data from the SCADA network to business networks. However, they must be designed and implemented properly to avoid introduction of additional risk through improper configuration.
3. **Evaluate and strengthen the security of any remaining connections to the SCADA networks.**
Conduct penetration testing or vulnerability analysis of any remaining connections to the SCADA network to evaluate the protection posture associated with these pathways. Use this information in conjunction with risk management processes to develop a robust protection strategy for any pathways to the SCADA network. Since the SCADA network is only as secure as its weakest connecting point, it is essential to implement firewalls, intrusion detection systems (IDSs), and other appropriate security measures at each point of entry. Configure firewall rules to prohibit access from and to the SCADA network, and be as specific as possible when permitting approved connections. For example, an Independent System Operator (ISO) should not be granted "blanket" network access simply because there is a need for a connection to certain components of

the SCADA system. Strategically place IDSs at each entry point to alert security personnel of potential breaches of network security. Organization management must understand and accept responsibility or risks associated with any connection to the SCADA network.

4. **Harden SCADA networks by removing or disabling unnecessary services.**
SCADA control servers built on commercial or open-source operating systems can be exposed to attack default network services. To the greatest degree possible, remove or disable unused services and network demons to reduce the risk of direct attack. This is particularly important when SCADA networks are interconnected with other networks. Do not permit a service or feature on a SCADA network unless a thorough risk assessment of the consequences of allowing the service/feature shows that the benefits of the service/feature far outweigh the potential for vulnerability exploitation. Examples of services to remove from SCADA networks include automated meter reading/remote billing systems, email services, and Internet access. An example of a feature to disable is remote maintenance. Numerous secure configurations such as the National Security Agency's series of security guides. Additionally, work closely with SCADA vendors to identify secure configurations and coordinate any and all changes to operational systems to ensure that removing or disabling services does not cause downtime, interruption of service, or loss of support.
5. **Do not rely on proprietary protocols to protect your system.**
Some SCADA systems have unique, proprietary protocols for communications between field devices and servers. Often the security of SCADA systems is based solely on the secrecy of these protocols. Unfortunately, obscure protocols provide very little "real" security. Do not rely on proprietary protocols or factor default configuration setting to protect your system. Additionally, demand that vendors disclose any backdoors or vendor interfaces to your SCADA systems, and expect them to provide systems that are capable of being secured.
6. **Implement the security features provided by device and system vendors.**
Older SCADA systems (most systems in use) have no security features whatsoever. SCADA system owners must insist that their system vendor implement security features in the form of product patches or upgrades.

Some newer SCADA devices are shipped with basic security features, but these are usually disabled to ensure ease of installation.

Analyze each SCADA device to determine whether security features are present. Additionally, factory default security settings (such as in computer network firewalls) are often set to provide maximum usability but minimal security. Set all security features to provide the maximum security only after a thorough risk assessment of the consequences of reducing the security level.

7. **Establish strong controls over any medium that is used as a backdoor into the SCADA network.**
Where backdoors or vendor connections do exist in SCADA systems, strong authentication must be implemented to ensure secure communications. Modems, wireless, and wired networks used for communications and maintenance represent a significant vulnerability to the SCADA network and remote sites. Successful "war dialing" or "war driving" attacks could allow an attacker to bypass all of other controls and have direct access to the SCADA network or resources. To minimize the risk of such attacks, disable inbound access and replace it with some type of callback system.
8. **Implement internal and external intrusion detection systems and establish 24-hour-a-day incident monitoring.**
To be able to effectively respond to cyber attacks, establish an intrusion detection strategy that includes alerting network administrators of malicious network activity originating from internal or external sources. Intrusion detection system monitoring is essential 24 hours a day; this capability can be easily set up through a pager. Additionally, incident response procedures must be in place to allow an effective response to any attack. To complement network monitoring, enable logging on all systems and audit system logs daily to detect suspicious activity as soon as possible.
9. **Perform technical audits of SCADA devices and networks, and any other connected networks, to identify security concerns.**
Technical audits of SCADA devices and networks are critical to ongoing security effectiveness. Many commercial and open-sourced security tools are available that allow system administrators to conduct audits of their systems/networks to identify active services, patch level, and common

vulnerabilities. The use of these tools will not solve systemic problems, but will eliminate the "paths of least resistance" that an attacker could exploit. Analyze identified vulnerabilities to determine their significance, and take corrective actions as appropriate. Track corrective actions and analyze this information to identify trends. Additionally, retest systems after corrective actions have been taken to ensure that vulnerabilities were actually eliminated. Scan non-production environments actively to identify and address potential problems.

10. **Conduct physical security surveys and assess all remote sites connected to the SCADA network to evaluate their security.**
Any location that has a connection to the SCADA network is a target, especially unmanned or unguarded remote sites. Conduct a physical security survey and inventory access points at each facility that has a connection to the SCADA system. Identify and assess any source of information including remote telephone/computer network/fiber optic cables that could be tapped; radio and microwave links that are exploitable computer terminals that could be accessed; and wireless local area network access points. Identify and eliminate single points of failure. The security of the site must be adequate to detect or prevent unauthorized access. Do not allow "live" network access points at remote, unguarded sites simply for convenience.
11. **Establish SCADA "Red Teams" to identify and evaluate possible attack scenarios.**
Establish a "Red Team" to identify potential attack scenarios and evaluate potential system vulnerabilities. Use a variety of people who can provide insight into weaknesses of the overall network, SCADA system, physical systems, and security controls. People who work on the system every day have great insight into the vulnerabilities of your SCADA network and should be consulted when identifying potential attack scenarios and possible consequences. Also, ensure that the risk from a malicious insider is fully evaluated, given that this represents one of the greatest threats to an organization. Feed information resulting from the "Red Team" evaluation into risk management processes to assess the information and establish appropriate protection strategies.

The following steps focus on management actions to establish an effective cybersecurity program:

12. **Clearly define cybersecurity roles, responsibilities, and authorities for managers, system administrators, and users.**
Organization personnel need to understand the specific expectations associated with protecting information technology resources through the definition of clear and logical roles and responsibilities. In addition, key personnel need to be given sufficient authority to carry out their assigned responsibilities. Too often, cybersecurity is left up to the initiative of the individual, which usually leads to inconsistent implementations and ineffective security. Establish a cybersecurity organizational structure that defines roles and responsibilities and clearly identifies how cybersecurity issues are escalated and who is notified in an emergency.
13. **Document network architecture and identify systems that serve critical functions or contain sensitive information that require additional levels of protection.**
Develop and document robust information security architecture as part of a process to establish an effective protection strategy. It is essential that organizations design their network with security in mind and continue to have a strong understanding of their network architecture throughout its life cycle. Of particular importance, an in-depth understanding of the functions that the systems perform and the sensitivity of the stored information is required. Without this understanding, risk cannot be properly assessed and protection strategies may not be sufficient. Documenting the information security architecture and its components is critical to understanding the overall protection strategy, and identifying single points of failure.
14. **Establish a rigorous, ongoing risk management process.**
A thorough understanding of the risks to network computing resources from denial-of-service attacks and the vulnerability of sensitive information to compromise is essential to an effective cybersecurity program. Risk assessments from the technical basis of this understanding are critical to formulating effective strategies to mitigate vulnerabilities and

preserve the integrity of computing resources. Initially, perform a baseline risk analysis based on current threat assessment to use for developing a network protection strategy. Due to rapidly changing technology and the emergence of new threats on a daily basis, an ongoing risk assessment process is also needed so that routine changes can be made to the protection strategy to ensure it remains effective. Fundamental to risk management is identification of residual risk with a network protection strategy in place and acceptance of that risk by management.

15. **Establish a network protection strategy based on the principle of defense-in-depth.**
A fundamental principle that must be part of any network protection strategy is defense-in-depth. Defense-in-depth must be considered early in the design phase of the development process, and must be an integral consideration in all technical decision-making associated with the network. Utilize technical and administrative controls to mitigate threats from identified risks to as great a degree as possible at all levels of the network. Single points of failure must be avoided, and cybersecurity defense must be layered to limit and contain the impact of any security incidents. Additionally, each layer must be protected against other systems at the same layer. For example, to protect against the inside threat, restrict users to access only those resources necessary to perform their job functions.

16. **Clearly identity cybersecurity requirements.**
Organizations and companies need structured security programs with mandated requirements to establish expectations and allow personnel to be held accountable. Formalized policies and procedures are typically used to establish and institutionalize a cybersecurity program. A formal program is essential in establishing a consistent, standards-based approach to cybersecurity through an organization and eliminates sole dependence on individual initiative. Policies and procedures also inform employees of their specific cybersecurity responsibilities and the consequences of failing to meet those responsibilities. They also provide guidance regarding actions to be taken during a cybersecurity incident and promote efficient and effective actions during a time of crisis. As part of identifying cybersecurity requirements, include user agreements and notification and warning banners. Establish requirements to minimize the threat from malicious insiders, including the

need for conducting background checks and limiting network privileges to those absolutely necessary.

17. **Establish effective configuration management processes.**
A fundamental management process needed to maintain a secure network is configuration management. Configuration management needs to cover both hardware configurations and software configurations. Changes to hardware or software can easily introduce vulnerabilities that undermine network security. Processes are required to evaluate and control any change to ensure that the network remains secure. Configuration management begins with well-tested and documented security baselines for your various systems.

18. **Conduct routine self-assessments.**
Robust performance evaluation processes are needed to provide organizations with feedback on the effectiveness of cybersecurity policy and technical implementation. A sign of a mature organization is one that is able to self-identify issues, conduct root cause analyses, and implement effective corrective actions that address individual and systemic problems. Self-assessment processes that are normally part of an effective cybersecurity program include routine scanning for vulnerabilities, automated auditing of the network, and self-assessments of organizational and individual performance.

19. **Establish system backups and disaster recovery plans.**
Establish a disaster recovery plan that allows for rapid recovery from any emergency (including a cyber attack). System backups are an essential part of any plan and allow rapid reconstruction of the network. Routinely exercise disaster recovery plans to ensure that they work and that personnel are familiar with them. Make appropriate changes to disaster recovery plans based on lessons learned from exercises.

20. **Senior organizational leadership should establish expectations for cybersecurity performance and hold individuals accountable for their performance.**
Effective cybersecurity performance requires commitment and leadership from senior managers in the organization. It is essential that senior management establish an expectation for strong cybersecurity and communicate this to their subordinate managers throughout the organization. It is also essential that senior organizational leadership establish a structure for

implementation of a cybersecurity program. This structure will promote consistent implementation and the ability to sustain a strong cybersecurity program. It is then important for individuals to be held accountable for their performance as it relates to cybersecurity. This includes managers, system administrators, technicians, and users/operators.

21. **Establish policies and conduct training to minimize the likelihood that organizational personnel will inadvertently disclose sensitive information regarding SCADA system design, operations, or security controls.** Release data related to the SCADA network only on a strict, need-to-know basis, and only to persons explicitly authorized to receive such information. "Social engineering," the gathering of information about a computer or computer network via questions to naïve users, is often the first step in a malicious attack on computer networks. The more information revealed about a computer or computer network, the more vulnerable the computer/network is. Never divulge data revealed to a SCADA network, including the names and contact information about the system operators/administrators, computer operating systems, and/or physical and logical locations of computers and network systems over telephones or to personnel unless they are explicitly authorized to receive such information. Any requests for information by unknown persons need to be sent to a central network security location for verification and fulfillment. People can be a weak link in an otherwise secure network. Conduct training and information awareness campaigns to ensure that personnel remain diligent in guarding sensitive network information, particularly their passwords.

REFERENCES AND RECOMMENDED READING

Associated Press (AP). (2009). *Goal: Disrupt.* From the 04/04/09 *The Virginian-Pilot*, Norfolk, VA.

Brown, A. S. (2008). *SCADA vs. The Hackers.* American Society of Mechanical Engineers. Accessed 05/10/08. http://www.memagazine.org/backissues/dec02/features/scadavs/.

DOE. (2001). *21 Steps to Improve Cyber Security of SCADA Networks.* Washington, DC: Department of Energy.

Ezell, B. C. (1998). *Risks of Cyber Attack to Supervisory Control and Data Acquisition*. Charlottesville: University of Virginia.

FEMA. (2008). *FEMA452: Risk Assessment A How to Guide*. Accessed 05/01/08. fema.gov/library/file?type=published/filetofile.

FEMA. (2015). *Protecting Critical Infrastructure Against Insider Threats*. Accessed 17 April 2015. http://emilms.fema.gov/IS0915/IABsummary.htm.

FBI. (2000). *Threat to Critical Infrastructure*. Washington, DC: Federal Bureau of Investigation.

FBI. (2007). *Ninth Annual Computer Crime and Security Survey*. Washington, DC: Computer Crime Institute and Federal Bureau of Investigations.

FBI. (2014). *Protecting Critical Infrastructure and the Importance of Partnerships*. https://www.fib.gov/news/speeches/protecitng-critical-infrastruce-andt-the-importantce-o...

GAO. (2003). *Critical Infrastructure Protection: Challenges in Securing Control System*. Washington, DC: United States General Accounting Office.

GAO. (2015). *Critical Infrastructure Protection: Sector-Specific Agencies Need to Better Measure Cybersecurity Progress*. Washington, DC: United States Government Accountability Office.

Gellman, B. (2002). Cyber-Attacks by Al-Qaeda Feared: Terrorists at Threshold of Using Internet as Tool of Bloodshed, Experts Say. *Washington Post*, June 27; p. A01.

Minter, J. G. (1996). Prevention of Chemical Accidents Still A Challenge. *Occupational Hazards*, September.

National Infrastructure Advisory Council. (2008). *First Report and Recommendations on the Insider Threat to Critical Infrastructure*. Washington, DC.

NIPC. (2002). *National Infrastructure Protection Center Report*. Washington, DC: National Infrastructure Protection Center.

Spellman, F. R. (1997). *A Guide to Compliance for PSM/RMP*. Lancaster, PA: Technomic Publishing Company.

Stamp, J., et al. (2003). *Common Vulnerabilities in Critical Infrastructure Control Systems*, 2nd edition. Albuquerque: Sandia National Laboratories.

U.S. Department of Energy. (2010). *Energy Sector-Specific Plan: An Annex to the National Infrastructure Protection Plan*. Washington, DC: USDOE.

US DOE. (2002). *Vulnerability Assessment Methodology: Electric Power Infrastructure*. Washington, DC.

USEPA. (2005). EPA Needs to Determine What Barriers Prevent Water Systems from Securing Known SCADA Vulnerabilities. In Harris, J., *Final Briefing Report*. Washington, DC: USEPA.

Warwalking. (2003). Accessed 05/09/08. http://warwalking.tribe.net.

Wiles, J., et al. (2007). *Techno Security's™ Guide to Securing SCADA* . Burlington, MA: Elsevier, Inc.

Young, M. A. (2004). *SCADA Systems Security*. SANS Institute.

8

Emergency Response

We're in uncharted territory.

—Rudi Giuliani (9/11/01)

When New York Mayor Rudi Giuliani made the above statement to Police Commissioner Bernard Kerik at the World Trade Center Site, September 11, 2001, to a point and to a degree, one of the first (and not to be forgotten) gross understatements of the twenty-first century had been uttered. Indeed, for citizens of the United States of America, the 9/11 events placed our level of consciousness, awareness, fear, and questions of what to do next in "uncharted territory." Actually, when you get right down to it, 9/11 generated more questions than anything else. Many are still asking the following questions today:

Why?

Why would anyone have the audacity to attack the U.S.?

What kind of cold-blooded killers would even think of conducting such an event?

Who were those Islamic radicals who perpetrated 9/11?

What were the terrorists' goal(s)?

Why?

Why were we not ready for such an attack?

Why had we not foreseen such an event?

Why were our emergency responders so undermanned, ill-prepared, and ill-equipped to handle such a disaster?

What took the military fighter planes so long to respond?

What did our government really know (if anything) before the events occurred?

Could anyone have prevented it?

Bottom line questions: Why us? Hell, why anyone?

Why?

These and several other questions continue to resonate today; no doubt they will continue to haunt us for some time to come.

Maybe we ask post 9/11-related questions because of who we are, what we are, and what we are not. That is, because we are Americans we are free, uninhibited thinkers who think what we say and say what we think—isn't America great! Most Americans are soft-hearted and sympathetic to those in need—compassion is the very nature and soul of being American. Americans are not born terrorists; they are not born into a terrorist regime; they are not raised with fear in their hearts—they are not afraid every time they leave their homes and go about their daily business. Suicide bombers and other like terrorists are those that occupy some other faraway place, definitely not America, and they are definitely not American. Right?

Notwithstanding exceptions to the rule, such as Timothy McVeigh (a so-called red-blooded American, born and raised in America) and that other idiot (whether a national or foreigner) who mailed the anthrax, terrorism was foreign to us.

Today, from a safety/security point of view, based on the events of 9/11 and the anthrax events, we should no longer be asking why. Instead, we should not waste our time, money and energy asking why or in pointing a finger of blame at our government, military, 9/11 emergency responders, and/or the terrorists. We should stop asking why and shift our mindset to asking what-if. The point is we need to stop feeling sorry for ourselves and accept the fact that there are folks out there who do not share our view of the American way of life. Earlier, in regard to security preparedness, we pointed to the need to ask what-if questions. Simply, what-if analysis is a proactive approach used to prevent or mitigate certain disasters, extreme events—those human- or nature-generated. Obviously, asking and properly answering what-if questions has little affect on preventing the actions of Mother Nature, such as earthquakes, tornadoes, hurricanes (Katrina type-events), and others. On the other hand, it is true that what-if questions, when properly posed and answered (with results), can reduce the death toll and overall damage caused by these natural disasters. We are certainly aware that these natural events are possible, probable, and likely, and their effects can be horrendous—beyond tragic. The irony is apparent, however, especially when we ask: how many of us are actually willing to move away from or out of earthquake zones, hurricane and tornado allies, and floodplains to live somewhere else?

The fact is we do not possess a crystal ball to foretell the future. What-if questions prepare us to react and respond to certain contingencies. And respond we must, because there are certain events we simply can't prevent. The best response to an event we can't prevent is summed up by the Boy Scouts Motto: Be Prepared!

COMMUNICATIONS SECTOR CONTINGENCY PLANNING

Emergency response planning or contingency planning for extreme events has long been standard practice for emergency planners and safety professionals in industrial systems operations. For many years, prudent practices have required consideration of the potential impact of severe natural events (forces of nature), including earthquakes, tornadoes, volcanoes, floods, hurricanes, and blizzards. These possibilities have been included in communications sector industry infrastructure emergency preparedness and disaster response planning. In addition, many communications sector production and cyber distribution facilities have considered the potential consequences of man-caused disasters such as operator error and manufacturer's industrial equipment defects. Currently, communications sector managers and operators must also consider violence in the workplace. Moreover, at present, as this text has pointed out, there is a new focus of concern: the potential effects of intentional acts by domestic (homegrown—in-house) or international (foreign) terrorists.

As a result, the security paradigm has not necessarily changed, but instead has been radically adjusted—reasonable, necessary and sensible accommodation for and mitigation of just about any emergency situation imaginable have been and continue to be made. Because we cannot foresee all future domestic or intentional acts of terrorism, we must be prepared to shift from the proactive to reactive mode on short notice—in some cases, on very short notice. Accordingly, we must be prepared to respond to, react to, and mitigate what we can't prevent. Unfortunately, there is more we can't prevent than we can prevent. In light of this, in this section we present in outline form a reactive mitigation procedure, the template example for a standard Communication Sector Crisis Communications Plan, dealing with natural and man-caused disasters, which could also be applied in response to acts of terrorism.

CRISIS COMMUNICATIONS PLAN

Note: The following criteria have not been established as anything other than guidelines and are offered not as definitive or official regulations or procedures but rather as informed advice (based on more than 30 years of safety, industrial hygiene, emergency contingency planning and security experience) insofar as to the subject matter specific to both public and private sectors.

Note: This emergency action plan applies to locations and facilities that are occupied by communications personnel performing their designated work activities.

The fact is well known: When an emergency occurs, the need to communicate is immediate. The goals of a Crisis Communications Plan are to document and understand the steps needed to:

- Rapidly restore material producing, refining, transporting, generating, transmitting, conserving, building, distributing, maintaining, and controlling systems and components after an emergency.
- Minimize communications sector production process equipment damage.
- Minimize impact and loss to customers.
- Minimize negative impacts on public health and employee safety.
- Minimize adverse effects on the environment.
- Provide emergency public information concerning customer service.

- Provide hazardous chemical information for first responders and other outside agencies.

Although we are concerned with the Communications Sector in this text, the USEPA developed *Large Water System Emergency Response Outline: Guidance to Assist Community Water Systems in Complying with the Public Health and Bioterrorism Preparedness and Response Act of 2002* (dated July 2003), with minor adjustments, can be applied as a template for the communications sector. This template provides guidance and recommendations to aid facilities in the preparation of emergency response plans under the PL 107-188. The template is provided below.

Crisis Communications Plan Template

I. Introduction

Safe and reliable operation is vital to every industrial operation. Crisis Communications Plan (CCP) is an essential part of managing a communications sector plant, process, or entity. The introduction should identify the requirement to have a documented Crisis Communications Plan, the goal(s) of the plan (e.g., be able to quickly identify an emergency and initiate timely and effective response action, be able to quickly respond and repair damages to minimize system downtime), and how access to the plan is limited. Plans should be numbered for control. Recipients should sign and date a statement that includes their (1) CCP number, (2) agreement not to reproduce the CCP, and (3) that they have read the CCP.

CCPs do not necessarily need to be one document. They may consist of an overview document, individual Emergency Action Procedures, checklists, additions to existing operations manuals, appendices, and so forth. There may be separate, more detailed plans for specific incidents. There may be plans that do not include particularly sensitive information and those that do. Existing applicable documents should be referenced in the CCP.

II. Emergency Planning Process

A. Planning Partnerships

The planning process should include those parties who will need to help the communications sector in an emergency situation (e.g., first responders, law enforcement, public health officials, nearby utilities, local emergency planning committees, testing labs, etc.). Partnerships should track from the communications sector operation up through local, state, regional, and federal agencies, as applicable and appropriate, and could also document compliance with governmental requirements.

B. General Emergency Response Policies, Procedures, Actions, Documents

A short synopsis of the overall emergency management structure, how other industrial emergency response, contingency, and risk management plans fit into the Crisis Communication Plan for communications sector emergencies, and applicable policies, procedures, action plans, and reference documents should be cited. Policies should include interconnect agreements with adjacent communities and just how the Crisis Communications Plan may affect them.

C. Scenarios

Use your Vulnerability Assessment (VA) findings to identify specific emergency action steps required for response, recovery, and remediation for applicable incident types. In Section V of this plan, specific emergency action procedures addressing each of the incident types should be addressed.

III. Emergency Response Plan—Policies

A. System Specific Information

In an emergency, communications sector industries need to have basic information for system personnel and external parties such as law enforcement, emergency responders, repair contractors/vendors, the media, and others. The information needs to be clearly formatted and readily accessible so system staff can find and distribute it quickly to those who may be involved in responding to the emergency. Basic information that may be presented in the emergency response plan are the system's ID number, system name, system address or location, directions to the system, population served, number of service connections, system owner, and information about the person in charge of managing the emergency. Distribution maps, detailed plant drawings, site plans, source/storage/production energy locations, and operations manuals may be attached to this plan as appendices or referenced.

1. PWS ID, Owner, Contact Person
2. Population served and service connections
3. System Components
   a) Conduits and constructed conveyances
   b) Physical barriers
   c) Electronic, computer, or other automated systems which are utilized by the communications sector industry
   d) Emergency power generators (onsite and portable)
   e) The operation and maintenance of such system components

B. Chain-of-Command Chart developed in Coordination with Local Emergency Planning Committee (Internal and/or External Emergency Responders, or both)
1. Contact Name
2. Organization and Emergency Response Responsibility
3. Telephone number(s) (hardwire, cell phones, faxes, e-mail)
4. State 24-hour Emergency Communications Center Telephone

C. Communications Procedures: Who, What, When
During most emergencies, it will be necessary to quickly notify a variety of parties both internal and external to the communications sector entity. Using the Chain-of-Command Chart and all appropriate personnel from the lists below, indicate who activates the plan, the order in which notification occurs, and the members of the Emergency Response Team. All contact information should be available for routine updating and readily available. The following lists are not intended to be all inclusive—they should be adapted to your specific needs.
1. Internal Notification Lists
   a) Operations Dispatch
   b) Communications Sector Plant Manager
   c) Communications Sector Processing Manager
   d) Communications Production/Storage/Distribution Manager
   e) Facility Managers
   f) Chief Energy Engineer
   g) Director of Engineering
   h) Data (IT) Manager

   i) Maintenance Manager
   j) Other
2. Local Notification
   a) Head of local government (i.e., Mayor, City Manager, Chairman of Board, etc.)
   b) Public Safety Officials—Fire, Local Law Enforcement (LLE), Police, EMS, Safety if a malevolent act is suspected LLE should be immediately notified and in turn will notify the FBI, if required. The FBI is the primary agency for investigating sabotage.
   c) Other Government Entities: Health, Schools, Parks, Finance, Electric, etc.
3. External Notification Lists
   a) State Department of Environmental Quality (DEQ)
   b) USEPA/USDOE/DHS/FCC
   c) State Police
   d) State Health Department (lab)
   e) Critical Customers (Special considerations for hospitals, federal, state, and country government centers, etc).
   f) Service Aid
   g) Mutual Aid
   h) Commercial customers not previously notified
4. Public/Media Notification: When and How to Communicate Effective communications is a key element of emergency response, and a media or communications plan is essential to good communications. Be prepared by organizing basic facts about the crisis and your chemical system. Develop key messages to use with the media that are clear, brief, and accurate. Make sure your messages are carefully planned and have been coordinated with local and state officials. Considerations should be given to establishing protocols for both field and office staff to respectfully defer questions to the utility spokesperson.

   Be prepared to list geographic boundaries of the affected area, (e.g., west of highway a, east of highway b, north of highway c and south of highway d to ensure the public clearly understands the system boundaries.)

D. Personnel Safety

This should provide direction as to how operations staff, emergency responders, and the public should respond to a potential toxic chemical release, including facility evacuation, personnel accountability, proper Personnel Protective Equipment as dictated by the Risk Management Program and Process Safety Management Plan, and whether the nearby public should be "in-place sheltered" or evacuated.

E. Equipment

The ERP should identify equipment that can obviate or significantly lessen the impact of terrorist attacks or other intentional actions on the public health and protect the safety and supply of communities and individuals. The communications sector facility should maintain an updated inventory of current equipment and repair parts for normal maintenance work.

Because of the potential for extensive or catastrophic damage that could result from a malevolent act, additional equipment sources should be identified for the acquisition and installation of equipment and repair parts in excess of normal usage. A certain number of "long-lead" procurement equipment should be inventoried and the vendor information for such unique and critical equipment maintained. In addition, mutual aid agreements with other industries, and the equipment available under the agreement, should be addressed. Inventories of current equipment, repair parts, and associated vendors should be indicated.

F. Property Protection

A determination should be made as to what communication sector producing, processing, distribution operation/facility should be immediately "locked down," specific access control procedures implemented, initial security perimeter established, a possible secondary malevolent event considered. The initial act may be a divisionary act.

G. Training, Exercises, and Drills

Emergency response training is essential. The purpose of the training program is to inform employees of what is expected of them during an emergency situation. The level of training on CCP directly affects how well a communications sector facility's employees can respond to an emergency. This may take the form of orientation scenarios, table-top workshops, functional exercises, etc.

H. Assessment

To evaluate the overall CCP's effectiveness and to ensure that procedures and practices developed under the CCP are adequate and are being implemented, communications sector industry staff should audit the program on a periodic basis.

IV. Emergency Action Procedures (EAPs)

These are detailed procedures used in the event of an operation emergency or malevolent act. EAPs may be applicable across many different emergencies and are typically common core elements of the overall municipality Emergency Response Plan (ERP) (e.g., responsibilities, notifications lists, security procedures, etc.) and can be referenced.

A. Event classification/severity of emergency
B. Responsibilities of emergency director
C. Responsibilities of incident commander
D. Emergency Operations Center (EOC) activation
E. Division internal communications and reporting
F. External communications and notifications
G. Emergency telephone list (division internal contacts)
H. Emergency telephone list (off-site responders, agencies, state 24-hour emergency phone number, and others to be notified)
I. Mutual aid agreements
J. Contact list of available emergency contractor services/equipment
K. Emergency equipment list (including inventory for each facility)
L. Security and access control during emergencies
M. Facility evacuation and lockdown and personnel accountability
N. Treatment and transport of injured personnel (including electrocution and petrochemical exposure)
O. Petrochemical records—to compare against historical results for base line
P. List of available labs for emergency use
Q. Emergency sampling and analysis (petrochemical)

R. Water use restrictions during emergencies
S. Alternate temporary chemical supplies during emergencies
T. Isolation plans for chemical supply, treatment, storage, and distribution systems
U. Mitigation plans for neutralizing, flushing, and collecting spilled chemicals
V. Protection of vital records during emergencies
W. Record keeping and reporting (FCC, FEMA, DHS, DOT, OSHA, EPA, and other requirements) (it is important to maintain accurate financial records of expenses associated with the emergency event for possible federal reimbursement.)
X. Emergency program training, drills/and tabletop exercises
Y. Assessment of emergency management plan and procedures
Z. Crime scene preservation training and plans
AA. Communication plans:
   1. Police
   2. Fire
   3. Local government
   4. Media
   5. And so forth
BB. Administration and logistics, including EOC, when established
CC. Equipment needs/maintenance of equipment
DD. Recovery and restoration of operations
EE. Emergency event closeout and recovery

V. Incident-Specific Emergency Action Procedures (EAPs)

Incident-Specific EAPs are action procedures that identify specific steps in responding to an operational emergency or malevolent act.

A. General response to terrorist threats (other than bomb threat and incident-specific threats)
B. Incident-specific response to man-made or technological emergencies
   1. Contamination event (articulated threat with unspecified materials)
   2. Contamination threat at a major event
   3. Notification from health officials of potential contamination

4. Intrusion through Supervisory Control and Data Acquisition (SCADA)

C. Significant structural damage resulting from intentional act
D. Customer complaints
E. Severe weather response (snow, ice, temperature, lightning)
F. Flood response
G. Hurricane and/or tornado response
H. Fire response
I. Explosion response
J. Major vehicle accident response
K. Electrical power outage response
L. Water supply interruption response
M. Transportation accident response—barge, plane, train, semi-trailer/tanker
N. Contaminated/tampered with water treatment chemicals
O. Earthquakes response
P. Disgruntled employees response (i.e., workplace violence)
Q. Vandals response
R. Bomb threat response
S. Civil disturbance/riot/strike
T. Armed intruder response
U. Suspicious mail handling and reporting
V. Hazardous chemical spill release response (including material safety data sheets)
W. Cybersecurity/Supervisory Control and Data Acquisition (SCADA) system attack response (other than incident specific, e.g., hacker)

VI. Next Steps

A. Plan review and approval
B. Practice and Plan to Update (as necessary, once every year recommended)

1. Training requirements
2. Who is responsible for conducting training, exercises, and emergency drills?
3. Update and assessment requirements
4. Incident-specific requirements

VII. Annexes
   A. Facility and Location Information
      1. Facility maps
      2. Facility drawings
      3. Facility descriptions/layout
      4. Etc.
VIII. References and Links
   A. Department of Homeland Security—http://www.dhs/gov/dhspublic
   B. Environmental Protection Agency—http://www.epa.gov
   C. Federal Emergency Management Agency—http://ww.fema.gov
   D. Local Emergency Planning Committees—http://www.epa.gov/ceppo/epclist.htm

THE BOTTOM LINE

Because industrial emergencies (in less than extreme conditions) can seriously affect the surrounding community and environment, and because poor planning and/or panic can only make a bad situation worse, and can also lead to additional injury and death, your role as communications sector site manager or site safety professional in an emergency response is doubly important. A crisis out of hand can easily devastate a community—and your organization is (or should be) an active member of your community. By ensuring less than effective emergency response, communications site managers endanger not only themselves and their organizations but also endanger their organization's community and standing as well.

REFERENCES AND RECOMMENDED READING

Brauer, R. L. (1994). *Safety and Health for Engineers*. New York: Van Nostrand Reinhold.

CoVan, J. (1995). *Safety Engineering*. New York: John Wiley and Sons.

DHS. (2010). Communications Sector-Specific Plan. Washington, DC: U.S. Department of Homeland Security.

Healy, R. J. (1969). *Emergency and Disaster Planning*. New York: Wiley.

Office of the Federal Register, 29 CFR 1910.120. Washington, DC: Office of the Federal Register, 1987.

Planning Guide and Checklist for Hazardous Materials Contingency Plans. FEMA-10, Federal Emergency Management Agency, Washington, DC, July 1981.

Safety and Health Requirements Manual, rev. ed. EM 385-1-1, U.S. Army Corps of Engineers, Washington, DC, October 1987.

Smith, A. J. (1980). *Managing Hazardous Substances Accidents*. New York: McGraw-Hill.

Spellman, F. R. (1997). *A Guide to Compliance for Process Safety Management Planning (PSM/RMP)*. Lancaster, PA: Technomic Publishing Company.

USDOE. (2008). *Emergency Support Function #12—Energy Annex*. Washington, DC: U.S. Department of Energy.

U.S. Department of Energy. (2010). *Energy Sector-Specific Plan: An Annex to the National Infrastructure Protection Plan*. Washington, DC: USDOE.

USEPA. (2002). *Water Utility Response, Recovery & Remediation Guidance for Man-made and/or Technological Emergencies*. Washington, DC: United States Environmental Protection Agency.

USEPA. (2003). *Large Water System Emergency Response Plan Outline: Guidance to Assist Community Water Systems in Complying with the Public Health Security and Bioterrorism Preparedness and Response Act of 2002*. EPA 810-F-03-007. Accessed 6/06. www.epa.gov/safewater/security. Washington, DC: United States Environmental Protection Agency.

9

Security Techniques and Hardware

> There are but two powers in the world, the sword and the mind. In the long run the sword is always beaten by the mind.
>
> —Napoleon Bonaparte

If you facility still uses keyed locks, does anyone in your facility control those keys? Does anyone in your facility know who has keys? Does anyone know how many keys have been issued? Is there a recorded record of each key that has been issued?

I know your password . . . therefore, you belong to me.

THE MULTIPLE-BARRIER APPROACH

Ideally, in a perfect world, all communication sector physical sites/facilities/towers would be secured in a layered fashion (aka the multiple barrier approach). Layered security systems are vital. Using the protection "in depth" principle, requiring that an adversary defeat several protective barriers or security layers to accomplish its goal, communications sector physical infrastructure can be made more secure. Protection in depth is a term commonly used by the military to describe security measures that reinforce one another, masking the defense mechanisms from the view of intruders, and allowing the defender time to respond to intrusion or attack.

A prime example of the use of the multi-barrier approach to ensure security and safety is demonstrated by the practices of the bottled water industry. In the aftermath of 9/11 and the increased emphasis on homeland security, a shifted paradigm of national security and vulnerability awareness has emerged. Recall that in the immediate aftermath of the 9/11 tragedies, emergency responders and others responded quickly and worked to exhaustion. In addition to the emergency responders, bottled water companies responded immediately by donating several million bottles of water to the crews at the crash sites in New York, at the Pentagon, and in Pennsylvania. International Bottled Water Association (IBWA; 2004) reports that "within hours of the first attack, bottled water was delivered where it mattered most—to emergency personnel on the scene who required ample water to stay hydrated as they worked to rescue victims and clean up debris" (p. 2).

Bottled water companies continued to provide bottled water to responders and rescuers at the 9/11 sites throughout the post-event(s) process(es). These patriotic actions by the bottled water companies, however, beg the question: How do we ensure the safety and security of the bottled water provided to anyone? IBWA (2004) has the answer: Using a multi-barrier approach, along with other principles, will enhance the safety and security of bottled water. IBWA (2004) describes its multi-barrier approach as follows:

> **A multi-barrier approach**—Bottled water products are produced utilizing a multi-barrier approach, from source to finished product, that helps prevent possible harmful contaminants (physical, chemical or microbiological) from adulterating the finished product as well as storage, production, and transportation equipment. Measures in a multi-barrier approach may include source protection, source monitoring, reverse osmosis, distillation, filtration, ozonation or ultraviolet (UV) light. Many of the steps in a multi-barrier system may be effective in safeguarding bottled water from microbiological and other contamination. Piping in and out of plants, as well as storage silos and water tankers are also protected and maintained through sanitation procedures. In addition, bottled water products are bottled in a controlled, sanitary environment to prevent contamination during the filling operation. (p. 3)

In communications sector infrastructure security, protection in depth is used to describe a layered security approach. A protection in depth strategy uses several forms of security techniques and/or devices against an intruder and does not rely on one single defensive mechanism to protect infrastructure. By implementing multiple layers of security, a hole or flaw in one layer is covered by the other layers. An intruder will have to intrude through each layer without being detected in the process—the layered approach implies that no matter how an intruder attempts to accomplish his goal, he will encounter effective elements of the physical protection system.

For example, as depicted in Figure 9.1, an effective security layering approach requires that an adversary penetrate multiple, separate barriers to gain entry to a critical TARGET at a communications sector facility. As shown in Figure 9.1, protection in depth (multiple layers of security) helps to ensure

FIGURE 9.1
Layered approach to security.

that the security system remains effective in the event of a failure or an intruder bypassing a single layer of security.

Again, as shown in Figure 9.1, layered security starts with the outer perimeter (the fence—the first line of physical security) of the facility and goes inward to the facility, the buildings, structures, other individual assets, and finally to the contents of those buildings—the TARGETs.

The area between the outer perimeter and structures or buildings is known as the site. This open site area provides an incomparable opportunity for early identification of an unauthorized intruder and initiation of early warning/response. This open space area is commonly used to calculate the standoff distance: that is, it is the distance between the outside perimeter (public areas to the fence) and the TARGET or critical assets (buildings/structures) inside the perimeter (inside the fence line—the restricted access area).

The open area, between the perimeter fence and the Target (e.g., operations center), if properly outfitted with various security devices, can also provide layered protection against intruders. For example, lighting is a deterrent. Based on personal experience, an open area within the plant site that is almost as well lighted at night as would be expected during daylight hours is the rule of thumb. In addition, strategically placed motion detectors along with crash barriers at perimeter gate openings and in front of vital structures are also recommended. Armed, mobile guards who roam the interior of the plant site on a regular basis provide the ultimate in site area security.

The next layer of physical security is the outside wall of the target structure(s) itself. Notwithstanding door, window and/or skylight entry, walls prevent most intruders from easy entry. If doors can only be entered using card reader access, security is shored up or enhanced to an extent. The same can be said for windows and skylights that are fashioned small enough to prohibit normal human entry. These same "weak" spots in buildings can be bastioned with break-proof or reinforced security glass.

The final layer of security is provided by properly designed interior features of buildings. Examples of these types of features include internal doors and walls, equipment cages, and backup or redundant equipment.

In the preceding discussion, the conditions described referred to perfect world conditions—that is, to those conditions that we "would want" (i.e., the security manager's proverbial wish list) to be incorporated into the design and installation of new communications sector infrastructure.

Post–9/11, in a not so perfect world, however, many of the peripheral (fence line) measures described above are more difficult to incorporate into communications sector infrastructure. This is not to say that communications sector sites and facilities do not have fence lines or fences; many of them do. These fences are designed to keep vandals, thieves and trespassers out. The problem is the fact that many of these facilities were constructed several years ago before urban encroachment literally encircled the many sites, allowing, at present, little room for security stand backs or setbacks to be incorporated into electrical power stations, plants and/or critical equipment locations. Based on personal observation, many of these fences face busy city streets or closely abut structures outside the fence line. The point is that when one sits down to plan a security upgrade, these factors must be taken into account.

Managers of communications sector infrastructure have four primary security areas to manage. These security areas are listed and described below.

- Physical Security—in the communications sector physical security techniques and practices have the most effect at "fenced" locations. At such locations, a systems approach is best, where detection, assessment, communication, and response are planned and supported by resources, procedures, and policies.
- Cyber/Information Technology Security—only the use of SCADA and other key operating systems that have been properly vetted and scrubbed of alleged Chinese and/or Russian Trojan Horses hacked into the North American electrical grid is important. The only positive way to ensure the security of the North American grid is to disconnect its cyber and other digital systems from the Internet. This step is impractical at the present time, but points to the need to conduct frequent audits of the system and install firewall protection in SCADA and other systems to prevent hacking. Frequent third-party penetration testing is advised.
- Employment Screening—mitigates the threat from the enemy at the water cooler (inside the organization). We are always amazed whenever we conduct security audits for various companies. Often, a simple check such as reviewing an employee's driving record often reveals that the employee has no license, is driving on a suspended license, or has a horrific driving record. Hiring standards and pre-employment background investigations

may help ensure the trustworthiness and reliability of personnel who have unescorted access to critical facilities.

- Protecting Potentially Sensitive Information—the old saying that goes—a secret can best be held between three people so long as two of the three are dead—makes the point that reducing the likelihood that information could be used by those intent on disrupting operations or causing death or destruction in communications sector plants/sites is crucial. Information should only be shared within an organization on a need-to-know basis.

For existing facilities, security upgrades should be based on the results generated from the vulnerability assessment, which characterizes and prioritizes those assets that may be targeted. Those vulnerabilities identified must be protected.

In the following sections, various security hardware and/or devices are described. These devices serve the main purpose of providing security against physical and/or digital intrusion. That is, they are designed to delay and deny intrusion and are normally coupled with detection and assessment technology. Possible additional security measures, based on the vulnerability assessment that may be recommended (covered in this text) include the following (NAERC 2002):

- Electronic security
- Closing non-essential perimeter and internal portals
- Physical barriers such as bollards or Jersey walls
- Fencing
- Lighting
- Security surveys
- Vulnerability assessments
- Availability of security resources
- General personnel and security officer training
- Law enforcement liaison
- Ensure availability of essential spare parts (machines, repair parts, wire, pipe, valves, transformers, etc.) for critical facilities

Keep in mind, however, and as mentioned previously, no matter the type of security device or system employed, communications sector systems cannot be made immune to all possible intrusions or attacks. Whenever a plant or facility safety/security manager tells us that he or she has secured their site 100 percent, we are reminded of Schneier's (2000) view of security: "You can't defend. You can't prevent. The only thing you can do is detect and respond." Simply, when it comes to making "anything" absolutely secure from intrusion or attack, there is inherently, or otherwise, no silver bullet."

In the next section, security hardware devices are discussed in detail. Before we describe these hardware devices, keep in mind that in addition to security hardware devices to help protect and monitor communications sector assets, there are a few employee practices and actions that can be taken to protect communications assets. For example, when a company computer system fails and must be disposed of, how is it disposed of? Is there a procedure or practice in place to prevent the valuable information on the system's hard drive from being pulled from a trash heap or from a dumpster and used by potential enemies? Are shredders used? Are they state of the art shredders that prevent scraps from being reassembled by enemies? Are building cleaning crews properly vetted and supervised? Do you have a team that routinely inspects suspended ceilings for bugs, cameras and listening recorders? Do you have key stroke reader capability; that is, can you record what messages are being sent by employees? Do you routinely check hard-wired phone lines? Have you removed all door signs that tell anyone what is on the other side of the door? Have you trained your employees to be slightly suspicious of just about anything and everything? Have you opened your manholes lately to see what is inside?

All of these practices just mentioned do not require security hardware such as barriers, motion detectors, fences, locks, biometric systems, video cameras, armed guards, electrified fences, and so forth. What they require instead is common sense, awareness and alert and engaged supervisors and employees. The point is what is heard at the water cooler is sometimes more significant than any security alarm apparatus can provide.

SECURITY HARDWARE DEVICES[1]

USEPA (2005) groups infrastructure security devices or products described below into four general categories:

1. It is important to point out that even though the following USEPA security asset and device recommendations were first made for the water/wastewater critical infrastructure, these recommendations are applicable to all other critical infrastructure sectors, including the communications sector.

- Physical asset monitoring and control devices
- Cyber protection devices
- Communication/integration
- Environmental monitoring devices

Physical Asset Monitoring and Control Devices

Aboveground, Outdoor Equipment Enclosures

Communications sector facilities and sites can consist of multiple structural components spread over a wide area and typically include a centralized production and distribution center, as well as component storage facilities that are distributed at multiple locations throughout the area. One of the primary reasons for constructing structural components that house operational equipment aboveground eliminates the safety risks associated with confined space entry, which is often required for the maintenance of equipment located below ground. In addition, space restrictions often limit the amount of equipment that can be located inside, and there are concerns that some types of equipment (such as backflow prevention devices—to prevent chemicals and fuel wastes from entering plant and offsite potable water systems) can, under certain circumstances, discharge fuel slurry or waste mixtures that could flood pits, vaults, or equipment rooms. In regard to electrical power, electrical substations are not usually suited for underground installation. Therefore, many pieces of critical electrical manufacturing equipment are located outdoors and aboveground in configurations that are properly fenced, insulated or isolated to prevent accidental electrical shock or short circuits/fires in equipment.

Experience demonstrates that many different system components can be and are often installed outdoors and aboveground, many of them controlled by wireless communication devices. Examples of these types of components could include:

- Backflow prevention devices
- Air release and control valves
- Pressure vacuum breakers
- Oil and gas pumps and motors
- Petrochemical storage and feed equipment

- Meters
- Sampling equipment
- Instrumentation
- Electrical substations
- Oil and natural gas pipelines

One of the most effective security measures for protecting aboveground equipment, where feasible, is to place it inside a building or exterior fenced structure. When/where this is not possible, enclosing the equipment or parts of the equipment using some sort of commercial or homemade add-on structure may help to prevent tampering with the equipment. These types of add-on structures or enclosures, which are designed to protect people and animals from electrocution and to protect equipment both from the elements and from unauthorized access or tampering, typically consist of a box-like fenced structure that is placed over or around the entire component, or over/around critical parts of the component (i.e., valves, etc.), and is then secured to delay or prevent intruders from tampering with the equipment. The enclosures are typically locked or otherwise anchored to a solid foundation, which makes it difficult for unauthorized personnel to remove the enclosure and access the equipment.

Standardized aboveground enclosures are available in a wide variety of materials, sizes, and configurations. Many options and security features are also available for each type of enclosure, and this allows system operators the flexibility to customize an enclosure for a specific application and/or price range. In addition, most manufacturers can custom-design enclosures if standard, off-the-shelf enclosures do not meet a user's needs.

Many of these enclosures are designed to meet certain standards. For example, the American Society of Sanitary Engineers (ASSE) has developed Standard #1060, *Performance Requirements for Outdoor Enclosures for Backflow Prevention Assemblies.* If an enclosure will be used to house a backflow preventer, this standard specifies the acceptable construction materials for the enclosure, as well as the performance requirements that the enclosure should meet, including specifications for freeze protection, drainage, air inlets, access for maintenance, and hinge requirements. ASSE #1060 also states that the enclosure should be lockable to enhance security.

Electrical substation and electrical equipment enclosures must meet the requirements and recommendations of various OHSA Standards, National Fire Protection Association (NFPA), National Electrical Codes (NEC), Institute of Electrical and Electronic Engineers (IEEE), and local code requirements.

Equipment enclosures can generally be categorized into one of four main configurations, which include:

- One piece, drop-over enclosures
- Hinged or removable top enclosures
- Sectional enclosures
- Shelters with access locks

All enclosures, including those with integral floors, must be secured to a foundation to prevent them from being moved or removed. Poorly anchored or unanchored enclosures may be blown off the equipment being protected, or may be defeated by intruders. In either case, this may result in the equipment beneath the enclosure becoming exposed and damaged. Therefore, ensuring that the enclosure is securely anchored will increase the security of the protected equipment.

The three basic types of foundations that can be used to anchor the aboveground equipment enclosure are concrete footers, concrete slab-on-grade, or manufactured fiberglass pads. The most common types of foundations utilized for equipment enclosures are standard or slab-on-grade footers; however, local climate and soil conditions may dictate whether either of these types of foundations can be used. These foundations can be either precast or poured in place at the installation site. Once the foundation is installed and properly cured, the equipment enclosure is bolted or anchored to the foundation to secure it in place.

An alternative foundation, specifically for use with smaller Hot Box® enclosures, is a manufactured fiberglass pad known as the Glass Pad™. The Glass Pad™ has the center cut out so that it can be dropped directly over the piece of equipment being enclosed. Once the pad is set level on the ground, it is backfilled over a two-inch flange located around its base. The enclosure is then placed on top of the foundation, and is locked in place with either a staple or slotted anchor, depending on the enclosure configuration.

One of the primary attributes of a security enclosure is its strength and resistance to breaking and penetration. Accordingly, the materials from which the enclosure is constructed will be important in determining the strength of the enclosure, and thus its usefulness for security applications. Enclosures are typically manufactured for either fiberglass or aluminum. With the exception of the one-piece, drop-over enclosure, which is typically fabricated from fiberglass, each configuration described above can be constructed from either material. In addition, enclosures can be custom-manufactured from polyurethane, galvanized steel, or stainless steel. Galvanized or stainless steel is often offered as an exterior layer, or "skin," for an aluminum enclosure. Although they are typically utilized in underground applications, precast concrete structures can also be used as aboveground equipment enclosures. However, precast structures are much heavier and more difficult to maneuver than are their fiberglass and aluminum counterparts. Concrete is also brittle, and that can be a security concern; however, products can be applied to concrete structures to add strength and minimize security risks (i.e., epoxy coating). Because precast concrete structures can be purchased from any concrete producers, this document does not identify specific vendors for these types of products.

In addition to the construction materials, enclosure walls can be configured or reinforced to give them added strength. Adding insulation is one option that can strengthen the structural characteristics of an enclosure; however, some manufacturers offer additional features to add strength to exterior walls. For example, while most enclosures are fabricated with a flat wall construction, some vendors manufacture fiberglass shelters with ribbed exterior walls. These ribs increase the structural integrity of the wall and allow the fabrication of standard shelters up to twenty feet in length. Another vendor has developed a proprietary process that uses a series of integrated fiberglass beams that are placed throughout a foam inner core to tie together the interior and exterior walls and roof. Yet another vendor constructs aluminum enclosures with horizontal and vertical redwood beams for structural support.

Other security features that can be implemented on aboveground, outdoor equipment enclosures include locks, mounting brackets, tamper-resistant doors, and exterior lighting.

Active Security Barriers (Crash Barriers)

Active security barriers (also known as crash barriers) are large structures that are placed in roadways at entrance and exit points to protected facilities to control vehicle access to these areas. These barriers are placed perpendicular to traffic to block the roadway, so that the only way that traffic can pass the barrier is for the barrier to be moved out of the roadway. These types of barriers are typically constructed from sturdy materials, such as concrete or steel, such that vehicles cannot penetrate through them. They are also designed at a certain height off the roadway so that vehicles cannot go over them.

The key difference between active security barriers, which include wedges, crash beams, gates, retractable bollards, and portable barricades, and passive security barriers, which include non-moveable bollards, jersey barriers, and planters, is that active security barriers are designed so that they can be raised and lowered or moved out of the roadway easily to allow authorized vehicles to pass them. Many of these types of barriers are designed so that they can be opened and closed automatically (i.e., mechanized gates, hydraulic wedge barriers), while others are easy to open and close manually (swing crash beams, manual gates). In contrast to active barriers, passive barriers are permanent, non-movable barriers, and thus they are typically used to protect the perimeter of a protected facility, such as sidewalks and other areas that do not require vehicular traffic to pass them. Several of the major types of active security barriers such as wedge barriers, crash beams, gates, bollards, and portable/removable barricades are described below.

Wedge barriers are plated, rectangular steel buttresses approximately 2-3 feet high that can be raised and lowered from the roadway. When they are in the open position, they are flush with the roadway and vehicles can pass over them. However, when they are in the closed (armed) position, they project up from the road at a 45-degree angle, with the upper end pointing toward the oncoming vehicle and the base of the barrier away from the vehicle. Generally, wedge barriers are constructed from heavy gauge steel, or concrete that contains an impact-dampening iron rebar core that is strong and resistant to breaking or cracking, thereby allowing them to withstand the impact from a vehicle attempting to crash through them. In addition, both of these materials help to transfer the energy of the impact over the barrier's entire volume, thus helping to prevent the barrier from being sheared off its base. In addition, because the barrier is angled away from traffic, the force of any vehicle impacting the barrier is distributed over the entire surface of the barrier and is not concentrated at the base, which helps prevent the barrier from breaking off at the base. Finally, the angle of the barrier helps hang up any vehicles attempting to drive over it.

Wedge barriers can be fixed or portable. Fixed wedge barriers can be mounted on the surface of the roadway ("surface-mounted wedges") or in a shallow mount in the road's surface, or they can be installed completely below the road surface. Surface-mounted wedge barriers operate by rising from a flat position on the surface of the roadway, while shallow-mount wedge barriers rise from their resting position just below the road surface. In contrast, below-surface wedge barriers operate by rising from beneath the road surface. Both the shallow-mounted and surface-mounted barriers require little or no excavation, and thus do not interfere with buried utilities. All three barrier mounting types project above the road surface and block traffic when they are raised into the armed position. Once they are disarmed and lowered, they are flush with the road, thereby allowing traffic to pass. Portable wedge barriers are moved into place on wheels that are removed after the barrier has been set into place.

Installing rising wedge barriers requires preparation of the road surface. Installing surface-mounted wedges does not require that the road be excavated; however, the road surface must be intact and strong enough to allow the bolts anchoring the wedge to the road surface to attach properly. Shallow-mount and below-surface wedge barriers require excavation of a pit that is large enough to accommodate the wedge structure, as well as any arming/disarming mechanisms. Generally, the bottom of the excavation pit is lined with gravel to allow for drainage. Areas not sheltered from rain or surface runoff can have a gravity drain or self-priming pump installed. Table 9.1 lists the pros and cons of wedge barriers.

Table 9.1. Pros and Cons of Wedge Barriers

Pros	*Cons*
Can be surface-mounted or completely installed below the roadway surface.	Installations below the surface of the roadway will require construction that may interfere with buried utilities.
Wedge barriers have a quick response time (normally 3.5–10.5 seconds, but can be 1–3 seconds in emergency situations). Because emergency activation of the barrier causes more wear and tear on the system than does normal activation, it is recommended for use only in true emergency situations.	Regular maintenance is needed to keep wedge fully operational.

(*Continued*)

Table 9.1. (Continued)

Pros	*Cons*
Surface or shallow-mount wedge barriers can be utilized in locations with a high water table and/or corrosive soils.	Improper use of the system may result in authorized vehicles being hung up by the barrier and damaged. Guards must be trained to use the system properly to ensure that this does not happen. Safety technologies may also be installed to reduce the risk of the wedge activating under an authorized vehicle.
All three wedge barrier designs have a high crash rating, thereby allowing them to be employed for higher security applications.	
These types of barrier are extremely visible, which may deter potential intruders.	

Source: USEPA (2005).

Crash beam barriers consist of aluminum beams that can be opened or closed across the roadway. While there are several different crash beam designs, every crash beam system consists of an aluminum beam that is supported on each side by a solid footing or buttress, which is typically constructed from concrete, steel, or some other strong material. Beams typically contain an interior steel cable (at least one inch in diameter) to give the beam added strength and rigidity. The beam is connected by a heavy-duty hinge or other mechanism to one of the footings so that it can swing or rotate out of the roadway when it is open, and can swing back across the road when it is in the closed (armed) position, blocking the road and inhibiting access by unauthorized vehicles. The non-hinged end of the beam can be locked into its footing, thus providing anchoring for the beam on both sides of the road and increasing the beam's resistance to any vehicles attempting to penetrate through it. In addition, if the crash beam is hit by a vehicle, the aluminum beam transfers the impact energy to the interior cable, which in turn transfers the impact energy through the footings and into their foundation, thereby minimizing the chance that the impact will snap the beam and allow the intruding vehicle to pass through.

Crash beam barriers can employ drop-arm, cantilever, or swing-beam designs. Drop-arm crash beams operate by raising and lowering the beam vertically across the road. Cantilever crash beams are projecting structures that are opened and closed by extending the beam from the hinge buttress to the receiving buttress located on the opposite side of the road. In the swing-beam design, the beam is hinged to the buttress such that it swings horizontally across the road. Generally, swing-beam and cantilever designs are used at locations where a vertical lift beam is impractical. For example, the swing-beam or cantilever designs are utilized at entrances and exits with overhangs, trees, or buildings that would physically block the operation of the drop-arm beam design.

Installing any of these crash beam barriers involves the excavation of a pit approximately 48 inches deep for both the hinge and the receiver footings. Due to the depth of excavation, the site should be inspected for underground utilities before digging begins. Table 9.2 lists the pros and cons of crash beams.

Table 9.2. Pros and Cons of Crash Beams

Pros	*Cons*
Requires little maintenance, while providing long-term durability.	Crash beams have a slower response time (normally 9.5–15.3 seconds, but can be reduced to 7–10 seconds in emergency situations) than do other types of active security barriers, such as wedge barriers. Because emergency activation of the barrier causes more wear and tear on the system than does normal activation, it is recommended for use only in true emergency situations.
No excavation is required in the roadway itself to install crash beams.	All three crash beam designs possess a low crash rating relative to other types of barriers, such as wedge barriers, and thus they typically are used for lower security applications.
	Certain crash barriers may not be visible to oncoming traffic and therefore may require additional lighting and/or other warning markings to reduce the potential for traffic to accidentally run into the beam.

Source: USEPA (2005).

In contrast to wedge barriers and crash beams, which are typically installed separately from a fence line, *gates* are often integrated units of a perimeter fence or wall around a facility.

Gates are basically movable pieces of fencing that can be opened and closed across a road. When the gate is in the closed (armed) position, the leaves of the gate lock into steel buttresses that are embedded in concrete foundations located on both sides of the roadway, thereby blocking access to the roadway. Generally, gate barricades are constructed from a combination of heavy gauge steel and aluminum that can absorb an impact from vehicles attempting to ram through them. Any remaining impact energy not absorbed by the gate material is transferred to the steel buttresses and their concrete foundations.

Gates can utilize a cantilever, linear, or swing design. Cantilever gates are projecting structures that operate by extending the gate from the hinge footing across the roadway to the receiver footing. A linear gate is designed to slide across the road on tracks via a rack and pinion drive mechanism. Swing gates are hinged so that they can swing horizontally across the road.

Installation of the cantilever, linear, or swing gate designs described above involve the excavation of a pit approximately 48 inches deep for both the hinge and receiver footings to which the gates are attached. Due to the depth of excavation, the site should be inspected for underground utilities before digging begins. Table 9.3 lists the pros and cons of gates.

Table 9.3. Pros and Cons of Gates

Pros	*Cons*
All three gate designs possess an intermediate crash rating, thereby allowing them to be utilized for medium to higher security applications.	Gates have a slower response time (normally 10–15 seconds, but can be reduced to 7–10 seconds in emergency situations) than do other types of active security barriers, such as wedge barriers. Because emergency activation of the barrier causes more wear and tear on the system than does normal activation, it is recommended for use only in true emergency situations.
Requires very little maintenance.	
Can be tailored to blend in with perimeter fencing.	
Gate construction requires no roadway excavation.	
Cantilever gates are useful for roads with high crowns or drainage gutters.	
These types of barriers are extremely visible, which may deter intruders.	
Gates can also be used to control pedestrian traffic.	

Source: USEPA (2005).

Bollards are vertical barriers at least 3 feet tall and 0.4 to 2 feet in diameter that are typically set 4 to 5 feet apart from each other so that they block vehicles from passing between them (see Figure 9.2). Smaller bollards, usually 4-inch diameter pipe filled with concrete, are installed in parking areas to prevent vehicles from striking walls or windows or to protect walkway areas. Bollards can either be fixed in place, removable, or retractable. Fixed and removable bollards are passive barriers that are typically used along building perimeters or on sidewalks to block vehicles from them, while allowing pedestrians to pass them. In contrast to passive bollards, retractable bollards are active security barriers that can easily be raised and lowered to allow vehicles to pass between them. Thus, they can be used in driveways or on roads to control vehicular access. When the bollards are raised, they protect

FIGURE 9.2
Bollard sequence placed in front of store fronts and pedestrian sidewalks in order to protect store property and pedestrians from rogue terrorist drivers using vehicles as weapons.
Source: Illustration by F. R. Spellman and Kathern Welsh.

the above the road surface and block the roadway; when they are lowered, they sit flush with the road surface, and thus allow traffic to pass over them. Retractable bollards are typically constructed from steel or other materials that have a low weight-to-volume ratio so that they require low power to raise and lower. Steel is also more resistant to breaking than is a more brittle material such as concrete, and is better able to withstand direct vehicular impact without breaking apart.

Retractable bollards are installed in a trench dug across a roadway—typically at an entrance or gate. Installing retractable bollards requires preparing the road surface. Depending on the vendor, bollards can be installed either in a continuous slab of concrete, or in individual excavations with concrete poured in place. The required excavation for a bollard is typically slightly wider and slightly deeper than the bollard height when extended aboveground. The bottom of the excavation is typically lined with gravel to allow drainage. The bollards are then connected to a control panel which controls the raising and lowering of the bollards. Installation typically requires mechanical, electrical, and concrete work; if utility personnel with these skills are available, then the utility can install the bollards themselves. Table 9.4 lists the pros and cons of retractable bollards.

Table 9.4. Pros and Cons of Retractable Bollards

Pros	*Cons*
Bollards have a quick response time (normally 3 to 10 seconds, but can be reduced to 1–3 seconds in emergency situations).	Bollard installations will require construction below the surface of the roadway, which may interfere with buried utilities.
Bollards have an intermediate crash rating, which allows them to be utilized for medium to higher security applications.	Some maintenance is needed to ensure barrier is free to move up and down.
	The distance between bollards must be decreased (i.e., more bollards must be installed along the same perimeter) to make these systems effective against small vehicles (i.e., motorcycles).

Source: USEPA (2005).

Portable/removable barriers, which can include removable crash beams and wedge barriers, are mobile obstacles that can be moved in and out of position on a roadway. For example, a crash beam may be completely removed and stored off-site when it is not needed. An additional example would be wedge barriers that are equipped with wheels that can be removed after the barricade is towed into place.

When portable barricades are needed, they can be moved into position rapidly. To provide them with added strength and stability, they are typically anchored to buttress boxes that are located on either side of the road. These buttress boxes, which may or may not be permanent, are usually filled with sand, water, cement, gravel, or concrete to make them heavy and aid in stabilizing the portable barrier. In addition, these buttresses can help dissipate any impact energy from vehicles crashing into the barrier itself.

Because these barriers are not anchored into the roadway, they do not require excavation or other related construction for installation. In contrast, they can be assembled and made operational in a short period of time. The primary shortcoming of this type of design is that these barriers may move if they are hit by vehicles. Therefore, it is important to carefully assess the placement and anchoring of these types of barriers to ensure that they can withstand the types of impacts that may be anticipated at that location. Table 9.5 lists the pros and cons of portable/removable barricades.

Table 9.5. Pros and Cons of Portable/Removable Barricades

Pros	*Cons*
Installing portable barricades requires no foundation or roadway excavation.	Portable barriers may move slightly when hit by a vehicle, resulting in a lower crash resistance.
Can be moved in and out of position in a short period of time.	Portable barricades typically require 7.75 to 16.25 seconds to move into place, and thus they are considered to have a medium response time when compared with other active barriers.
Wedge barriers equipped with wheels can be easily towed into place.	
Minimal maintenance is needed to keep barriers fully operational.	

Source: USEPA (2005).

Because the primary threat to active security barriers is that vehicles will attempt to crash through them, their most important attributes are their size, strength, and crash resistance. Other important features for an active security barrier are the mechanisms by which the barrier is raised and lowered to allow authorized vehicle entry, and other factors, such as weather resistance and safety features.

Alarms

An *alarm system* is a type of electronic monitoring system that is used to detect and respond to specific types of events—such as unauthorized access to an asset, or a possible fire. In chemical processing systems, alarms are also used to alert operators when process operating or monitoring conditions go out of preset parameters (i.e., process alarms). These types of alarms are primarily integrated with process monitoring and reporting systems (i.e., SCADA systems). Note that this discussion does not focus on alarm systems that are not related to a facility's processes.

Alarm systems can be integrated with fire detection systems, intrusion detection systems (IDSs), access control systems, or closed circuit television (CCTV) systems, such that these systems automatically respond when the alarm is triggered. For example, a smoke detector alarm can be set up to automatically notify the fire department when smoke is detected; or an intrusion alarm can automatically trigger cameras to turn on in a remote location so that personnel can monitor that location.

An alarm system consists of sensors that detect different types of events: an arming station that is used to turn the system on and off; a control panel that receives information, processes it, and transmits the alarm; and an annunciator that generates a visual and/or audible response to the alarm. When a sensor is tripped it sends a signal to a control panel, which triggers a visual or audible alarm and/or notifies a central monitoring station. A more complete description of each of the components of an alarm system is provided below.

Detection devices (also called *sensors*), are designed to detect a specific type of event (such as smoke, intrusion, etc.). Depending on the type of event they are designed to detect, sensors can be located inside or outside the facility or other asset. When an event is detected, the sensors use some type of communication method (such as wireless radio transmitters, conductors, or cables) to send signals to the control panel to generate the alarm. For example, a smoke detector sends a signal to a control panel when it detects smoke.

Alarms use either normally closed (NC) or normally open (NO) electric loops, or "circuits," to generate alarm signals. These two types of circuits are discussed separately below.

In NC loops or circuits, all of the system's sensors and switches are connected in a series. The contacts are "at rest" in the closed (on) position, and current continually passes through the system. However, when an event triggers the sensor, the loop is opened, breaking the flow of current through the system and triggering the alarm. NC switches are used more often than are NO switches because the alarm will be activated if the loop or circuit is broken or cut, thereby reducing the potential for circumventing the alarm. This is known as a "supervised" system.

In NO loops or circuits, all of the system's sensors and switches are connected in parallel. The contacts are "at rest" in the open (off) position, and no current passes through the system. However, when an event triggers the sensor, the loop is closed. This allows current to flow through the loop, powering the alarm. NO systems are not "supervised" because the alarm will not be activated if the loop or circuit is broken or cut. However, adding an end-of-line resistor to an NO loop will cause the system to alarm if tampering is detected.

An *arming station*, which is the main user interface with the security system, allows the user to arm (turn on), disarm (turn off), and communicate with the system. How a specific system is armed will depend on how it is used. For example, while IDSs can be armed for continuous operation (24 hours/day), they are usually armed and disarmed according to the work schedule at a specific location so that personnel going about their daily activities do not set off the alarms. In contrast, fire protection systems are typically armed 24hours/day.

A *control panel* receives information from the sensors and sends it to an appropriate location, such as to a central operations station or to a 24-hour monitoring facility. Once the alarm signal is received at the central monitoring location, personnel monitoring for alarms can respond (such as by sending security teams to investigate or by dispatching the fire department).

An *annunciator* responds to the detection of an event by emitting a signal. This signal may be visual, audible, electronic, or a combination of these three. For example, fire alarm signals will always be connected to audible annunciators, whereas intrusion alarms may not be.

Alarms can be reported locally, remotely, or both locally and remotely. Local and remotely (centrally) reported alarms are discussed in more detail below.

A *local alarm* emits a signal at the location of the event (typically using a bell or siren). A "local only" alarm emits a signal at the location of the event but does not transmit the alarm signal to any other location (i.e., it does not transmit the alarm to a central monitoring location). Typically, the purpose of a "local only" alarm is to frighten away intruders, and possibly to attract the attention of someone who might notify the proper authorities. Because no signal is sent to a central monitoring location, personnel can only respond to a local alarm if they are in the area and can hear and/or see the alarm signal.

Fire alarm systems must have local alarms, including both audible and visual signals. Most fire alarm signal and response requirements are codified in the National Fire Alarm Code, National Fire Protection Association (NFPA) 72. NFPA 72 discusses the application, installation, performance, and maintenance of protective signaling systems and their components. In contrast to fire alarms, which require a local signal when fire is detected, many IDSs do not have a local alert device, because monitoring personnel do not wish to inform potential intruders that they have been detected. Instead, these types of systems silently alert monitoring personnel that an intrusion has been detected, thus allowing monitoring personnel to respond.

In contrast to systems that are set up to transmit "local only" alarms when the sensors are triggered, systems can also be set up to transmit signals to a *central location*, such as to a control room or guard post at the utility, or to a police or fire station. Most fire/smoke alarms are set up to signal both at the location of the event and at a fire station or central monitoring station. Many insurance companies require that facilities install certified systems that include alarm communication to a central station. For example, systems certified by the Underwriters Laboratory (UL) require that the alarm be reported to a central monitoring station.

The main differences between alarm systems lie in the types of event detection devices used in different systems. Intrusion sensors, for example, consist of two main categories: perimeter sensors and interior (space) sensors. *Perimeter intrusion sensors* are typically applied on fences, doors, walls, or windows, and are designed to detect an intruder before he/she accesses a protected asset (i.e., perimeter intrusion sensors are used to detect intruders attempting to enter through a door, window, etc.). In contrast, *interior intrusion sensors* are designed to detect an intruder who has already accessed the protected asset (i.e., interior intrusion sensors are used to detect intruders once they are already within a protected room or building). These two types of detection devices can be complementary, and they are often used together to enhance security for an asset. For example, a typical intrusion alarm system might employ a perimeter glass-break detector that protects against intruders accessing a room through a window, as well as an ultrasonic interior sensor that detects intruders that have gotten into the room without using the window. Table 9.6 lists and describes types of perimeter and interior sensors.

Table 9.6. Perimeter and Interior Sensors

Type of Perimeter Sensor	*Description*
Foil	Foil is a thin, fragile, lead-based metallic tape that is applied to glass windows and doors. The tape is applied to the window or door, and electric wiring connects this tape to a control panel. The tape functions as a conductor and completes the electric circuit with the control panel. When an intruder breaks the door or window, the fragile foil breaks, opening the circuit and triggering an alarm condition.
Magnetic switches (reed switches)	The most widely used perimeter sensor. They are typically used to protect doors, as well as windows that can be opened (windows that cannot be opened are more typically protected by foil alarms).
Glass break detectors	Placed on glass and sense vibrations in the glass when it is disturbed. The two most common types of glass-break detectors are shock sensors and audio discriminators.
Type of Interior Sensor	*Description*
Passive infrared (PIR)	Presently the most popular and cost effective interior sensors. PIR detectors monitor infrared radiation (energy in the form of heat) and detect rapid changes in temperature within a protected area. Because infrared radiation is emitted by all living things, these types of sensors can be very effective.
Quad PIRs	Consist of two dual-element sensors combined in one housing. Each sensor has a separate lens and a separate processing circuitry, which allows each lens to be set up to generate a different protection pattern
Ultrasonic detectors	Emit high-frequency sound waves, and sense movement in a protected area by sensing changes in these waves. The sensor emits sound waves that stabilize and set a baseline condition in the area to be protected. Any subsequent movement within the protected area by a would-be intruder will cause a change in these waves, thus creating an alarm condition.
Microwave detectors	Emit ultra- high-frequency radio waves, and the detector senses any changes in these waves as they are reflected throughout the protected space. Microwaves can penetrate through walls, and thus a unit placed in one location may be able to protect multiple rooms.
Dual technology devices	Incorporate two different types of sensor technology (such as PIR and microwave technology) together in one housing. When both technologies sense an intrusion, an alarm is triggered.

Source: USEPA (2005).

Fire Detection/Fire Alarm Systems consist of different types of fire detection devices and fire alarm systems available. These systems may detect fire, heat, smoke, or a combination of any of these. For example, a typical fire alarm system might consist of heat sensors, which are located throughout a facility and which detect high temperatures or a certain change in temperature over a fixed time period. A different system might be outfitted with both smoke and heat detection devices. A summary of several different types of fire/smoke/heat detection sensors is provided in Table 9.7.

Table 9.7. Fire/Smoke/Heat Detection Sensors

Detector Type	*Description*
Thermal detectors	Sense when temperatures exceed a set threshold (fixed temperature detectors) or when the rate of change of temperature increases over a fixed time period (rate-of-rise detectors).
Duct detectors	Is located within the heating and ventilation ducts of the facility. This sensor detects the presence of smoke within the system's return or supply ducts. A sampling tube can be added to the detector to help span the width of the duct.
Smoke detectors	Sense invisible and/or visible products of combustion. The two principle types of smoke detectors are photoelectric and ionization detectors. The major differences between these devices are described below: • Photoelectric smoke detectors react to visible particles of smoke. These detectors are more sensitive to the cooler smoke with large smoke particles that is typical of smoldering fires. • Ionization smoke detectors are sensitive to the presence of ions produced by the chemical reactions that take place with few smoke particles, such as those typically produced by fast burning/flaming fires.
Multi-sensor detectors	Are a combination of photoelectric and thermal detectors. The photoelectric sensor serves to detect smoldering fires, while the thermal detector senses the heat given off from fast burning/flaming fires.
Carbon monoxide (CO) detectors	Are used to indicate the outbreak of fire by sensing the level of carbon monoxide in the air. The detector has an electrochemical cell which senses carbon monoxide, but not some other products of combustion.

Detector Type	Description
Beam detectors	Are designed to protect large, open spaces such as industrial warehouses. These detectors consist of three parts: the transmitter, which projects a beam of infrared light; the receiver, which registers the light and produces an electrical signal; and the interface, which processes the signal and generates an alarm of fault signals. In the event of a fire, smoke particles obstruct the beam of light. Once a preset threshold is exceeded, the detector will go into alarm.
Flame detectors	Sense either ultraviolet (UV) or infrared (IR) radiation emitted by a fire.
Air-sampling detectors	Actively and continuously sample the air from a protected space and are able to sense the pre-combustion stages of incipient fire.

Source: USEPA (2005).

Once a sensor in an alarm system detects an event, it must communicate an alarm signal. The two basic types of alarm communication systems are hardwired and wireless. Hardwired systems rely on wire that is run from the control panel to each of the detection devices and annunciators. Wireless systems transmit signals from a transmitter to a receiver through the air—primarily using radio or other waves. Hardwired systems are usually lower-cost, more reliable (they are not affected by terrain or environmental factors), and significantly easier to troubleshoot than are wireless systems. However, a major disadvantage of hardwired systems is that it may not be possible to hardwire all locations (for example, it may be difficult to hardwire remote locations). In addition, running wires to their required locations can be both time consuming and costly. The major advantage to using wireless systems is that they can often be installed in areas where hardwired systems are not feasible. However, wireless components can be much more expensive when compared to hardwired systems. In addition, in the past, it has been difficult to perform self-diagnostics on wireless systems to confirm that they are communicating properly with the controller. Presently, the majority of wireless systems incorporate supervising circuitry, which allows the subscriber to know immediately if there is a problem with the system (such as a broken detection device or a low battery), or if a protected door or window has been left open.

Backflow Prevention Devices

Backflow prevention devices are designed to prevent backflow, which is the reversal of the normal and intended direction of water flow in a water system. Backflow is a potential problem in a petrochemical processing system because if incorrectly cross-connected to potable water it can spread contaminated water back through a distribution system. For example, backflow at uncontrolled cross connections (cross-connections are any actual or potential connection between the public water supply and a source of chemical contamination) or pollution can allow pollutants or contaminants to enter the potable water system. More specifically, backflow from private plumbing systems, industrial areas, hospitals, and other hazardous contaminant-containing systems, into public water mains and wells poses serious public health risks and security problems. Cross-contamination from private plumbing systems can contain biological hazards (such as bacteria or viruses) or toxic substances that can contaminate and sicken an entire population in the event of backflow. The majority of historical incidences of backflow have been accidental, but growing concern that contaminants could be intentionally backfed into a system is prompting increased awareness for private homes, businesses, industries, and areas most vulnerable to intentional strikes. Therefore, backflow prevention is a major tool for the protection of water systems.

Backflow may occur under two types of conditions: backpressure and backsiphonage. *Backpressure* is the reverse from normal flow direction within a piping system that is the result of the downstream pressure being higher than the supply pressure. These reductions in the supply pressure occur whenever the amount of water being used exceeds the amount of water supplied, such as during water line flushing, fire fighting, or breaks in water mains. *Backsiphonage* is the reverse from normal flow direction within a piping system that is caused by negative pressure in the supply piping (i.e., the reversal of normal flow in a system caused by a vacuum or partial vacuum within the water supply piping). Backsiphonage can occur where there is a high velocity in a pipe line; when there is a line repair or break that is lower than a service point; or when there is lowered main pressure due to high water withdrawal rate, such as during fire fighting or water main flushing.

To prevent backflow, various types of backflow preventers are appropriate for use. The primary types of backflow preventers are:

- Air Gap Drains
- Double Check Valves
- Reduced Pressure Principle Assemblies
- Pressure Vacuum Breakers

Biometric Security Systems

Biometrics involves measuring the unique physical characteristics or traits of the human body. In ancient times biometrics involved the judging of one's accent, body hair or face to determine friend or foe. Presently, it is well known that any aspect of the body that is measurably different from person to person—for example, fingerprints or eye characteristics—can serve as a unique biometric identifier for that individual. Biometric systems recognizing fingerprints, palm shape, eyes, face, voice, and signature comprise the bulk of the current biometric systems, but those that recognize other biological features do exist.

Biometric security systems use biometric technology combined with some type of locking mechanisms to control access to specific assets. In order to access an asset controlled by a biometric security system, an individual's biometric trait must be matched with an existing profile stored in a database. If there is a match between the two, the locking mechanisms (which could be a physical lock, such as at a doorway, an electronic lock, such as at a computer terminal, or some other type of lock) is disengaged, and the individual is given access to the asset.

A biometric security system is typically comprised of the following components:

- A sensor, which measures/records a biometric characteristic or trait
- A control panel, which serves as the connection point between various system components. The control panel communicates information back and forth between the sensor and the host computer, and controls access to the asset by engaging or disengaging the system lock based on internal logic and information from the host computer
- A host computer, which processes and stores the biometric trait in a database
- Specialized software, which compares an individual image taken by the sensor with a stored profile or profiles
- A locking mechanism which is controlled by the biometric system
- A power source to power the system.

Biometric Hand and Finger Geometry Recognition

Hand and finger geometry recognition is the process of identifying an individual through the unique "geometry" (shape, thickness, length, width, etc.) of that individual's hand or fingers. Hand geometry recognition has been employed since the early 1980s and is among the most widely used biometric technologies for controlling access to important assets. It is easy to install and use, and is appropriate for use in any location requiring use of two finger highly accurate, non-intrusion biometric security. For example, it is currently used in numerous workplaces, day care facilities, hospitals, universities, airports, refineries, and power plants.

A newer option within hand geometry recognition technology is finger geometry recognition (not to be confused with fingerprint recognition). Finger geometry recognition relies on the same scanning methods and technologies as does hand geometry recognition, but the scanner only scans two of the user's fingers, as opposed to his entire hand. Finger geometry recognition has been in commercial use since the mid-1990s and is mainly used in time and attendance applications (i.e., to track when individuals have entered and exited a location). To date the only large-scale commercial use of two finger geometry for controlling access is at Disney World, where season pass holders use the geometry of their index and middle finger to gain access to the facilities.

To use a hand or finger geometry unit, an individual presents his or her hand or fingers to the biometric unit for "scanning." The scanner consists of a Charged Coupled Device (CCD), which is essentially a high resolution digital camera; a reflective platen on which the hand is placed; and a mirror or mirrors that help capture different angles of the hand or fingers. The camera "scans" individual geometric characteristics of the hand or fingers by taking multiple images while the user's hand rests on the reflective platen. The camera also captures "depth," or three-dimensional information, through light reflected from the mirrors and the reflective platen. This live image is then compared to a "template" that was previously established for that individual when they were "enrolled" in the system. If the live scan of the individual matches the stored template, the individual is "verified," and is given access to that asset. Typically, verification takes about two seconds. In access control applications, the scanner is usually connected to some sort of electronic lock, which unlocks the door, turnstile, or other entry barrier when the user is verified. The user can then proceed through the entrance. In time and attendance applications, the time that an individual checks in and out of a location is stored for later use.

As discussed above, hand and finger geometry recognition systems can be used in several different types of applications, including access control and time and attendance tracking. While time and attendance tracking can be used for security, it is primarily used for operations and payroll purposes (i.e., clocking in and clocking out). In contrast, access control applications are more likely to be security related. Biometric systems are widely used for access control, and can be used on various types of assets, including entryways, computers, vehicles, and so forth. However, because of their size, hand/finger recognition systems are primarily used in entryway access control applications.

Biometric Overview-Iris Recognition

The iris, which is the colored or pigmented area of the eye surrounded by the sclera (the white portion of the eye), is a muscular membrane that controls the amount of light entering the eye by contracting or expanding the pupil (the dark center of the eye). The dense, unique patterns of connective tissue in the human iris were first noted in 1936, but it was not until 1994, when algorithms for iris recognition were created and patented, that commercial applications using biometric iris recognition began to be used extensively. There are now two vendors producing iris recognition technology, both the original developer of these algorithms, as well as a second company, which has developed and patented a different set of algorithms for iris recognition.

The iris is an ideal characteristic for identifying individuals because it is formed *in utero*, and its unique patterns stabilize around eight months after birth. No two irises are alike; neither an individual's right or left irises, nor the irises of identical twins. The iris is protected by the cornea (the clear covering over the eye), and therefore it is not subject to the aging or physical changes (and potential variation) that are common to some other biometric measures, such as the hand, fingerprints, and face. Although some limited changes can occur naturally over time, these changes generally occur in the iris's melanin and therefore affect only the eye's color, and not its unique patterns (in addition, because iris scanning uses only black and white images, color changes would not affect the scan anyway). Thus, barring specific injuries or certain rare surgeries directly affecting the iris, the iris's unique patterns remain relatively unchanged over an individual's lifetime.

Iris recognition systems employ a monochromatic or black and white video camera that uses both visible and near infrared light to take video of an individual's iris. Video is used rather than still photography as an extra security procedure. The video is used to confirm the normal continuous fluctuations of the pupil as the eye focuses, which ensures that the scan is of a living human being, and not a photograph or some other attempted hoax. A high resolution image of the iris is then captured or extracted from the video, using a device often referred to as a "frame grabber." The unique characteristics identified in this image are then converted into a numeric code, which is stored as a "template" for that user.

Card Identification/Access/Tracking Systems

A card reader system is a type of electronic identification system that is used to identify a card and then perform an action associated with that card. Depending on the system, the card may identify where a person is or where they were at a certain time; or it may authorize another action, such as disengaging a lock. For example, a security guard may use his card at card readers located throughout a facility to indicate that he has checked a certain location at a certain time. The reader will store the information and/or send it to a central location, where it can be checked later to ensure that the guard has patrolled the area. Other card reader systems can be associated with a lock, so that the cardholder must have their card read and accepted by the reader before the lock disengages.

A complete card reader system typically consists of the following components:

- Access cards that are carried by the user
- Card readers, which read the card signals and send the information to control units
- Control units, which control the response of the card reader to the card
- A power source

A "card" may be a typical card or another type of device, such as a key fob or wand. These cards store electronic information, which can range from a simple code (i.e., the alphanumeric code on a Proximity card) to individualized personal data (i.e., biometric data on a Smartcard). The card reader reads the information stored on the card and sends it to the control unit, which determines the appropriate action to take when a card is presented. For example, in a card access system the control unit compares the information on the card versus stored access authorization information to determine if the card holder is authorized to proceed through the door. If the information stored in the card reader system indicates that the key is authorized to allow entrance through the doorway, the system disengages the lock and the key holder can proceed through the door.

There are many different types of card reader systems on the market. The primary differences between card reader systems are differences in the way that data is encoded on the cards and in the way these data are transferred between the card and the card reader, and in the types of applications for which they are best suited. However, all card systems are similar in the way that the card reader and control unit interact to respond to the card.

While card readers are similar in the way that the card reader and control unit interact to control access, they are different in the way data is encoded on the cards and the way these data are transferred between the card and the card reader. There are several types of technologies available for card reader systems. These include:

- Proximity
- Wiegand
- Smartcard
- Magnetic Stripe
- Bar Code
- Infrared
- Barium Ferrite
- Hollerith
- Mixed Technologies

Table 9.8 below summarizes various aspects of card reader technologies. The determination for the level of security rate (low, moderate, or high) is based on the level of technology a given card reader system has and how simple it is to duplicate that technology, and thus bypass the security. Vulnerability ratings were based on whether the card reader can be damaged easily due to frequent use or difficult working conditions (i.e., weather conditions if the reader is located outside). Often this is influenced by the number of moving parts in the system—the more moving parts, the greater the system's potential susceptibility to damage. The life cycle rating is based on the durability of a given card reader system over its entire operational period. Systems requiring frequent physical contact between the reader and the card often have a shorter life cycle due to the wear and tear to which the equipment is exposed. For many card reader systems, the vulnerability rating and life cycle ratings have a reciprocal relationship. For instance, if a given system has a high vulnerability rating, it will almost always have a shorter life cycle.

Table 9.8. Card Reader Technology

Types of Card Readers	*Technology*	*Life Cycle*	*Vulnerability*	*Level of Security*
Proximity	Embedded radio frequency circuits encoded with unique information	Long	Virtually none	Moderate-high
Wiegand	Short lengths of small-diameter, special alloy wire with unique magnetic properties	Long	Low susceptibility to damage; high durability due to embedded wires	Moderate-expensive
Magnetic stripe	Electromagnetic charges to encode information on a piece of tape attached to back of card	Moderate	Moderately susceptible to damage due to frequency of use	Low-Moderate
Bar code	Series of narrow and wide bars and spaces	Short	High; easily damaged	Low
Hollerith	Holes punched in a plastic or paper card and read optically	Short	High; easily damaged from frequent use	Low
Infrared	An encoded shadow pattern within the card, read using an infrared scanner	Moderate	IR scanners are optical and thus, vulnerable to contamination	High
Barium ferrite	Uses small bits of magnetized barium ferrite, placed inside a plastic card. The polarity and location of the "spots" determines the coding	Moderate	Low susceptibility to damage; durable since spots are embedded in the material	Moderate-High
Smartcards	Patterns or series of narrow and wide bars and spaces	Short	High susceptibility to damage, low durability	Highest

Source: USEPA (2005).

Card reader technology can be implemented for facilities of any size and with any number of users. However, because individual systems vary in the complexity of their technology and in the level of security they can provide to a facility, individual users must determine the appropriate system for their needs. Some important features to consider when selecting a card reader system include:

- The technological sophistication and security level of the card system
- The size and security needs of the facility
- The frequency with which the card system will be used. For systems that will experience a high frequency of use it is important to consider a system that has a longer life cycle and lower vulnerability rating, thus making it more cost effective to implement.
- The conditions in which the system will be used (i.e., will it be used on the interior or exterior of buildings, does it require light or humidity controls, etc.). Most card reader systems can operate under normal environmental conditions, and therefore this would be a mitigating factor only in extreme conditions
- System costs

Exterior Intrusion-Buried Sensors

Buried sensors are electronic devices that are designed to detect potential intruders. The sensors are buried along the perimeters of sensitive assets and are able to detect intruder activity both above- and below-ground. Some of these systems are composed of individual, stand-alone sensor units, while other sensors consist of buried cables.

There are four types of buried sensors that rely on different types of triggers. These are pressure or seismic, magnetic field, ported coaxial cable, and fiber-optic cables. These four sensors are all covert and terrain-following, meaning they are hidden from view and follow the contour of the terrain. The four types of sensors are described in more detail below. Table 9.9 presents the distinctions between the four types of buried sensors.

Table 9.9. Types of Buried Sensors

Type	*Description*
Pressure or seismic	Responds to disturbances in the soil
Magnetic field	Responds to a change in the local magnetic field caused by the movement of nearby metallic material
Ported coaxial cables	Responds to motion of a material with a high dielectric constant or high conductivity near the cables
Fiber optic cables	Responds to a change in the shape of the fiber that can be sensed using sophisticated sensors and computer signal processing

Source: Adapted from Garcia, M. L., 2001.

Exterior Intrusion Sensors

An exterior intrusion sensor is a detection device that is used in an outdoor environment to detect intrusions into a protected area. These devices are designed to detect an intruder, and then communicate an alarm signal to an alarm system. The alarm system can respond to the intrusion in many different ways, such as by triggering an audible or visual alarm signal, or by sending an electronic signal to a central monitoring location that notifies security personnel of the intrusion.

Intrusion sensors can be used to protect many kinds of assets. Intrusion sensors that protect physical space are classified according to whether they protect indoor, or "interior" space (i.e., an entire building or room within a building), or outdoor, or "exterior" space (i.e., a fence line or perimeter). Interior intrusion sensors are designed to protect the interior space of a facility by detecting an intruder who is attempting to enter, or who has already entered a room or building. In contrast, exterior intrusion sensors are designed to detect an intrusion into a protected outdoor/exterior area. Exterior protected areas are typically arranged as zones or exclusion areas placed so that the intruder is detected early in the intrusion attempt before the intruder can gain access to more valuable assets (e.g., into a building located within the protected area). Early detection creates additional time for security forces to respond to the alarm.

Exterior intrusion sensors are classified according to how the sensor detects the intrusion within the protected area. The three classes of exterior sensor technology include:

- Buried line sensors
- Fence-associated sensors
- Freestanding sensors

(1) Buried-Line Sensors—As the name suggests, buried line sensors are sensors that are buried underground, and are designed to detect disturbances within the ground—such as disturbances caused by an intruder digging, crawling, walking, or running on the monitored ground. Because they sense ground disturbances, these types of sensors are able to detect intruder activity both on the surface and below ground. Individual types of exterior buried line sensors function in different ways, including by detecting motion, pressure, or vibrations within the protected ground, or by detecting changes in some type of field (e.g., magnetic field) that the sensors generate within the protected ground. Specific types of buried line sensors include pressure or seismic sensors, magnetic field sensors, ported coaxial cables, and fiber-optic cables. Details on each of these sensor types are provided below.

- *Buried-line pressure* or *seismic sensors* detect physical disturbances to the ground—such as vibrations or soil compression—caused by intruders walking, driving, digging, or otherwise physically contacting the protected ground. These sensors detect disturbances from all directions and, therefore, can protect an area radially outward from their location; however, because detection may weaken as a function of distance from the disturbance, choosing the correct burial depth from the design area will be crucial. In general, sensors buried at a shallow depth protect a relatively small area but have a high probability of detecting intrusion within that area, while sensors buried at a deeper depth protect a wider area but have a lower probability of detecting intrusion into that area.
- *Buried line magnetic field sensors* detect changes in a local magnetic field that are caused by the movement of metallic objects within that field. This type of sensor can detect ferric metal objects worn or carried by an intruder entering a protected area on foot as well as vehicles being driven into the protected area.

- *Buried line ported coaxial cable sensors* detect the motion of any object (i.e., human body, metal, etc.) possessing high conductivity and located within close proximity to the cables. An intruder entering into the protected space creates an active disturbance in the electric field, thereby triggering an alarm condition.
- *Buried line fiber-optic cable sensors* detect changes in the attenuation of light signals transmitted within the cable. When the soil around the cable is compressed, the cable is distorted, and the light signal transmitted through the cable changes, initiating an alarm. This type of sensor is easy to install because it can be buried at a shallow burial depth (only a few centimeters) and still be effective.

(2) Fence-Associated Sensors—Fence-associated sensors are either attached to an existing fence, or are installed in such a way as to create a fence. These sensors detect disturbances to the fence—such as those caused by an intruder attempting to climb the fence, or by an intruder attempting to cut or lift the fence fabric. Exterior fence-associated sensors include fence-disturbance sensors, taut-wire sensor fences, and electric field or capacitance sensors. Details on each of these sensor types are provided below.

- *Fence-disturbance sensors* detect the motion or vibration of a fence that can be caused by an intruder attempting to climb or cut through the fence. In general, fence disturbance sensors are used on chain link fences or on other fence types where a moveable fence fabric is hung between fence posts.
- *Taut-wire sensor fences* are similar to fence-disturbance sensors except that instead of attaching the sensors to a loose fence fabric, the sensors are attached to a wire that is stretched tightly across the fence. These types of systems are designed to detect changes in the tension of the wire rather than vibrations in the fence fabric. Taut-wire sensor fences can be installed over existing fences, or as stand-alone fence systems.
- *Electric field or capacitance sensors* detect changes in capacitive coupling between wires that are attached to, but electrically isolated from, the fence. As opposed to other fence-associated intrusion sensors, both electric field and capacitance sensors generate an electric field that radiates out from the fence line, resulting in an expanded

zone of protection relative to other fence-associated sensors, and allowing the sensor to detect an intruders' presence before they arrive at the fence line. Note: proper spacing is necessary during installation of the electric field sensor to detect a would-be intruder from slipping between largely spaced wires.

(3) *Free-Standing Sensors*—These sensors, which include active infrared, passive infrared, bistatic microwave, monostatic microwave, dual-technology, and video motion detection (VMD) sensors, consist of individual sensor units or components that can be set up in a variety of configurations to meet a user's needs. They are installed above ground, and depending on how they are oriented relative to each other, they can be used to establish a protected perimeter or a protected space. More details on each of these sensor types are provided below.

- *Active infrared sensors* transmit infrared energy into the protected space, and monitor for changes in this energy caused by intruders entering that space. In a typical application, an infrared light beam is transmitted from a transmitter unit to a receiver unit. If an intruder crosses the beam, the beam is blocked, and the receiver unit detects a change in the amount of light received, triggering an alarm. Different sensors can see single- and multiple-beam arrays. Single-beam infrared sensors transmit a single infrared beam. In contrast, multiple-beam infrared sensors transmit two or more beams parallel to each other. This multiple-beam sensor arrangement creates an infrared "fence."
- *Passive infrared (PIR) sensors* monitor the ambient infrared energy in a protected area, and evaluate changes in that ambient energy that may be caused by intruders moving through the protected area. Detection ranges can exceed 100 yards on cold days with size and distance limitations dependent upon the background temperature. PIR sensors generate a non-uniform detection pattern (or "curtain") that has areas (or "zones") of more sensitivity and areas of less sensitivity. The specific shape of the protected area is determined by the detector's lenses. The general shape common too many detection patterns is a series of long "fingers" emanating from the PIR and spreading in various directions. When intruders enter the detection area, the PIR sensor detects differences in temperature due to the intruder's body heat, and triggers an alarm. While the

PIR leaves unprotected areas between its fingers, an intruder would be detected if he passed from a non-protected area to a protected area.

- *Microwave sensors* detect changes in received energy generated by the motion of an intruder entering into a protected area. Monostatic microwave sensors incorporate a transmitter and a receiver in one unit, while bistatic sensors separate the transmitter and the receiver into different units. Monostatic sensors are limited to a coverage area of 400 feet, while bistatic sensors can cover an area up to 1,500 feet. For bistatic sensors, a zone of no detection exists in the first few feet in front of the antennas. This distance from the antennas to the point at which the intruder is first detected is known as the offset distance. Due to this offset distance, antennas must be configured so that they overlap one another (as opposed to being adjacent to each other), thereby creating long perimeters with a continuous line of detection.
- *Dual-technology sensors* consist of two different sensor technologies incorporated into one sensor unit. For example, a dual technology sensor could consist of a passive infrared detector and a monostatic microwave sensor integrated into the same sensor unit.
- *Video motion detection* (VMD) *sensors* monitor video images from a protected area for changes in the images. Video cameras are used to detect unauthorized intrusion into the protected area by comparing the most recent image against a previously established one. Cameras can be installed on towers or other tall structures so that they can monitor a large area.

Fences

A fence is a physical barrier that can be set up around the perimeter of an asset. Fences often consist of individual pieces (such as individual pickets in a wooden fence, or individual sections of a wrought iron fence) that are fastened together. Individual sections of the fence are fastened together using posts, which are sunk into the ground to provide stability and strength for the sections of the fence hung between them. Gates are installed between individual sections of the fence to allow access inside the fenced area.

Many fences are used as decorative architectural features to separate physical spaces from each other. They may also be used to physically mark the location of a boundary (such as a fence installed along a property line). However, a fence can also serve as an effective means for physically delaying intruders from gaining access to a communications sector asset. For example, many utilities install fences around their primary facilities, around remote pump stations, or around hazardous petrochemical materials storage areas or sensitive areas within a facility. Access to the area can be controlled through security at gates or doors through the fence (for example, by posting a guard at the gate or by locking it). In order to gain access to the asset, unauthorized persons would either have to go around or through the fence.

Fences are often compared with walls when determining the appropriate system for perimeter security. While both fences and walls can provide adequate perimeter security, fences are often easier and less expensive to install than walls. However, they do not usually provide the same physical strength that walls do. In addition, many types of fences have gaps between the individual pieces that make up the fence (i.e., the spaces between chain links in a chain link fence or the spaces between pickets in a picket fence). Thus, many types of fences allow the interior of the fenced area to be seen. This may allow intruders to gather important information about the locations or defenses of vulnerable areas within the facility.

There are numerous types of materials used to construct fences, including chain link, iron, aluminum, wood, or wire. Some types of fences, such as split rails or pickets, may not be appropriate for security purposes because they are traditionally low fences, and they are not physically strong. Potential intruders may be able to easily defeat these fences either by jumping or climbing over them or by breaking through them. For example, the rails in a split fence can be broken easily.

Important security attributes of a fence include the height to which it can be constructed, the strength of the material comprising the fence, the method and strength of attaching the individual sections of the fence together at the posts and the fence's ability to restrict the view of the assets inside the fence. Additional considerations should include the ease of installing the fence and the ease of removing and reusing sections of the fence. Table 9.10 provides a comparison of the important security and usability features of various fence types.

Table 9.10. Comparison of Different Fence Types

Specifications	*Chain Link*	*Iron*	*Wire (Wirewall)*	*Wood*
Height limitations	12'	12'	12'	8'
Strength	Medium	High	High	Low
Installation requirements	Low	High	High	Low
Ability to remove/reuse	Low	High	Low	High
Ability to replace/repair	Medium	High	Low	High

Source: USEPA (2005).

Some fences can include additional measures to delay, or even detect, potential intruders. Such measures may include the addition of barbed wire, razor wire, or other deterrents at the top of the fence. Barbed wire is sometimes employed at the base of fences as well. This can impede an intruder's progress. Fences may also be fitted with security cameras to provide visual surveillance of the perimeter. Finally, some facilities have installed motion sensors along their fences to detect movement on the fence. Several manufacturers have combined these multiple perimeter security features into one product and offer alarms and other security features.

The correct implementation of a fence can make it a much more effective security measure. Security experts recommend the following when a facility constructs a fence:

- The fence should be at least 7–9 feet high.
- Any outriggers, such as barbed wire, that are affixed on top of the fence should be angled out and away from the facility, and not in toward the facility. This will make climbing the fence more difficult, and will prevent ladders from being placed against the fence.
- Other types of hardware can increase the security of the fence. This can include installing concertina wire along the fence (this can be done in front of the fence or at the top of the fence), or adding intrusion sensors, cameras, or other hardware to the fence.
- All undergrowth should be cleared for several feet (typically 6 feet) on both sides of the fence. This will allow for a clearer view of the fence by any patrols in the area.
- Any trees with limbs or branches hanging over the fence should be trimmed so that intruders cannot use them to go over the fence. Also, it should be noted that fallen trees can damage fences, and so management of trees

around the fence can be important. This can be especially important in areas where fence goes through a remote area.

- Fences that do not block the view from outside the fence to inside the fence allow patrols to see inside the fence without having to enter the facility.
- "No Trespassing" signs posted along a fence can be a valuable tool in prosecuting any intruders who claim that the fence was broken and that they did not enter through the fence illegally. Adding signs that highlight the local ordinances against trespassing can further persuade simple troublemakers from illegally jumping/climbing the fence. Electrical substation and other electrical component installations should have clearly visible signage warning of HIGH VOLTAGE and the dangers of Electrical Shock.

Films for Glass Shatter Protection

Many communications sector entities have numerous windows on the outside of buildings, in doors, and in interior offices. In addition, many facilities have glass doors or other glass structures, such as glass walls or display cases. These glass objects are potentially vulnerable to shattering when heavy objects are thrown or launched at them, when explosions occur near them, or when there are high winds (for exterior glass). If the glass is shattered, intruders may potentially enter an area. In addition, shattered glass projected into a room from an explosion or from an object being thrown through a door or window can injure and potentially incapacitate personnel in the room. Materials that prevent glass from shattering can help to maintain the integrity of the door, window, or other glass object, and can delay an intruder from gaining access. These materials can also prevent flying glass and thus reduce potential injuries.

Materials designed to prevent glass from shattering include specialized films and coatings. These materials can be applied to existing glass objects to improve their strength and their ability to resist shattering. The films have been tested against many scenarios that could result in glass breakage, including penetration by blunt objects, bullets, high winds, and simulated explosions. Thus, the films are tested against both simulated weather scenarios (which could include both the high winds themselves and the force of objects blown into the glass), as well as more criminal/terrorist scenarios where the glass is subject to explosives or bullets. Many vendors provide information on the results of these types of tests, and thus potential users can compare different product lines to determine which products best suit their needs.

The primary attributes of films for shatter protection are:

- The materials from which the film is made
- The adhesive that bonds the film to the glass surface
- The thickness of the film

Standard glass safety films are designed from high strength polyester. Polyester provides both strength and elasticity, which is important in absorbing the impact of an object, spreading the force of the impact over the entire film, and resisting tearing. The polyester is also designed to be resistant to scratching, which can result when films are cleaned with abrasives or other industrial cleaners.

The bonding adhesive is important in ensuring that the film does not tear away from the glass surface. This can be especially important when the glass is broken, so that the film does not peel off the glass and allow it to shatter. In addition, films applied to exterior windows can be subject to high concentrations of UV light, which can break down bonding materials.

Film thickness is measured in gauge or mils. According to test results reported by several manufacturers, film thickness appears to affect resistance to penetration/tearing, with thicker films being more resistant to penetration and tearing. However, the appreciation of a thicker film did not decrease glass fragmentation.

Many manufacturers offer films in different thicknesses. The "standard" film is usually one 4-mil layer; thicker films are typically composed of several layers of the standard 4-mil sheet. However, newer technologies have allowed the polyester to be "microlayered" to produce a stronger film without significantly increasing its thickness. In this microlayering process, each laminate film is composed of multiple micro-thin layers of polyester woven together at alternating angles. This provides increased strength for the film while maintaining the flexibility and thin profile of one film layer.

As described above, many vendors test their products in various scenarios that would lead to glass shattering, including simulated bomb blasts and simulation of the glass being struck by wind-blown debris. Some manufacturers refer to the Government Services Administration standard for bomb blasts, which require resistance to tearing for a 4-PSI blast. Other manufacturers use other measures and test for resistance to tearing. Many of these tests are not "standard," in that no standard testing or reporting methods have been adopted by any of the accepted standards-setting institutions. However, many of the vendors publish the procedure and the results of these tests on their websites, and this may allow users to evaluate the protectiveness of these films. For example, several vendors evaluate the "protectiveness" of their films and the "hazard" resulting from blasts near windows with and without protective films. Protectiveness is usually evaluated based on the percentage of glass ejected from the window, and the height at which that ejected glass travels during the blast (for example, if the blasted glass tends to project upward into a room—potentially toward people's faces—it is a higher hazard than if it is blown downward into the room toward people's feet). There are some standard measures of glass breakage. For example, several vendors indicated that their products exceed the American Society for Testing and Materials (ASTM) standard 64Z-95 "Standard Test Method for Glazing and Glazing Systems Subject to Air Blast Loadings." Vendors often compare the results of some sort of penetration or force test, ballistic tests, or simulated explosions with unprotected glass versus glass onto which their films have been applied. Results generally show that applying films to the glass surfaces reduces breakage/penetration of the glass and can reduce the amount and direction of glass ejected from the frame. This in turn reduces the hazard from flying glass.

In addition to these types of tests, many vendors conduct standard physical tests on their products, such as tests for tensile strength and peel strength. Tensile strength indicates the strength per area of material, while the peel strength indicates the force it would take to peel the product from the glass surface. Several vendors indicate that their products exceed American National Standards Institute (ANSI) standard Z97.1 for tensile strength and adhesion.

Vendors typically have a warranty against peeling or other forms of deterioration of their products. However, the warranty requires that the films be installed by manufacturer-certified technicians to ensure that they are applied correctly and therefore that the warranty is in effect. Warranties from different manufacturers may vary. Some may cover the cost of replacing the material only, while others include material plus installation. Because installation costs are significantly greater than material costs, different warranties may represent large differences in potential costs.

Fire Hydrant Locks

Fire hydrants are installed at strategic locations throughout a community's water distribution system to supply water for fire fighting. However, because there are many hydrants in a system and they are often located in residential neighborhoods, industrial districts, and other areas where they cannot be easily observed and/or guarded, they are potentially vulnerable to unauthorized access. Many municipalities, states, and EPA regions have recognized this potential vulnerability and have instituted programs to lock hydrants. For example, EPA Region 1 has included locking hydrants as number 7 on its "Drinking Water Security and Emergency Preparedness" Top Ten List for small groundwater suppliers.

A "hydrant lock" is a physical security device designed to prevent unauthorized access to the water supply through a hydrant. They can also ensure water and water pressure availability to fire fighters and prevent water theft and associated lost water revenue. These locks have been successfully used in numerous municipalities and in various climates and weather conditions.

Fire hydrant locks are basically steel covers or caps that are locked in place over the operating nut of a fire hydrant. The lock prevents unauthorized persons from accessing the operating nut and opening the fire hydrant valve. The lock also makes it more difficult to remove the bolts from the hydrant and access the system that way. Finally, hydrant locks shield the valve from being broken off. Should a vandal attempt to breach the hydrant lock by force and succeed in breaking the hydrant lock, the vandal will only succeed in bending the operating valve. If the hydrant's operating valve is bent, the hydrant will not be operational, but the water asset remains protected and inaccessible to vandals. However, the entire hydrant will need to be replaced.

Hydrant locks are designed so that the hydrants can be operated by special "key wrenches" without removing the lock. These specialized wrenches are generally distributed to the fire department, public works department, and other authorized persons so that they can access the hydrants as needed. An inventory of wrenches and their serial numbers is generally kept by a municipality so that the location of all wrenches is known. These operating key wrenches may only be purchased by registered lock owners.

The most important features of hydrants are their strength and the security of their locking systems. The locks must be strong so that they cannot be broken off. Hydrant locks are constructed from stainless or alloyed steel.

Stainless steel locks are stronger and are ideal for all climates; however, they are more expensive than alloy locks. The locking mechanisms for each fire hydrant locking system ensure that the hydrant can only be operated by authorized personnel who have the specialized key to work the hydrant.

Hatch Security

A hatch is basically a door that is installed on a horizontal plane (such as in a floor, a paved lot, or a ceiling), instead of on a vertical plane (such as in a building wall). Hatches are usually used to provide access to assets that are either located underground (such as hatches to basements or underground vaults and storage areas), or to assets located above ceilings (such as emergency roof exits). At communication facilities, hatches are typically used to provide access to underground vaults containing pumps, meter chambers, valves, or piping, or to the interior of chemical tanks or covered water reservoirs. Securing a hatch by locking it or upgrading materials to give the hatch added strength can help to delay unauthorized access to any asset behind the hatch.

Like all doors, a hatch consists of a frame anchored to the horizontal structure, a door or doors, hinges connecting the door/doors to the frame, and a latching or locking mechanism that keeps the hatch door/doors closed.

It should be noted that improving hatch security is straightforward, and that hatches with upgraded security features can be installed new or they can be retrofitted for existing applications.

Depending on the application, the primary security-related attributes of a hatch are the strength of the door and frame, its resistance to the elements and corrosion, it ability to be sealed against water or gas, and its locking features.

Hatches must be both strong and lightweight so that they can withstand typical static loads (such as people or vehicles walking or driving over them) while still being easy to open.

In addition, because hatches are typically installed at outdoor locations, they are usually designed from corrosion-resistant metal that can withstand the elements. Therefore, hatches are typically constructed from high gauge steel or lightweight aluminum.

Aluminum is typically the material of choice for hatches because it is lightweight and more corrosion resistant relative to steel. Aluminum is not as rigid as steel, so aluminum hatch doors may be reinforced with aluminum stiffeners to provide extra strength and rigidity. The doors are usually constructed from single or double layers (or "leaves") of material. Single-leaf designs are standard for smaller hatches, while double leaf deigns are required for larger hatches. In addition, aluminum products do not require painting. This is reflected in the warranties available with different products. Product warranties range from ten years to lifetime.

Steel is heavier per square foot than aluminum, and thus steel hatches will be heavier and more difficult to open than aluminum hatches of the same size. However, heavy steel hatch doors may have spring-loaded, hydraulic, or gas openers or other specialized features that help in opening the hatch and in keeping it open.

Many hatches are installed in outdoor areas, often in roadways or pedestrian areas. Therefore, the hatch installed for any given application must be designed to withstand the expected load at that location. Hatches are typically solid to withstand either pedestrian or vehicle loading. Pedestrian loading hatches are typically designed to withstand either 150 or 300 pounds per square feet (psf) of loading. The vehicle loading standard is the American Association of State Highway and Transportation Officials' (AASHTO) H-20 wheel loading standard of 16,000 pounds over an 8 inch by 20 inch area. It should be noted that these design parameters are for static loads and not dynamic loads; thus, the loading capabilities may not reflect potential resistance to other types of loads that may be more typical of an intentional threat, such as repeated blows from a sledge hammer or pressure generated by bomb blasts or bullets.

The typical design for a watertight hatch includes a channel frame that directs water away from the hatch. This can be especially important in a hatch on a storage tank because this will prevent liquid contaminants from being dumped on the hatch and leaking through into the interior. Hatches can also be constructed with gasket seals that are air-, odor-, and gas-tight.

Typically, hatches for pedestrian loading applications have hinges located on the exterior of the hatch, while hatches designed for H-20 loads have hinges located in the interior of the hatch. Hinges located on the exterior of the hatch may be removed, thereby allowing intruders to remove the hatch door and access the asset behind the hatch. Therefore, installing H-20 hatches even for applications which do not require H-20 loading levels may increase security, because intruders will not be able to tamper with the hinges and circumvent the hatch this way.

In addition to the location of the hinges, stock hinges can be replaced with heavy duty or security hinges that are more resistant to tampering.

The hatch locking mechanism is perhaps the most important part of hatch security. There are a number of locks that can be implemented for hatches, including:

- Slam locks (internal locks that are located within the hatch frame
- Recessed cylinder locks
- Bolt locks
- Padlocks

Ladder Access Control

Communications sector facilities have a number of assets that are raised above ground level, including electrical substations, transmitting stations, raised conduit systems, and roof access points into buildings. In addition, communications equipment, antennae, or other electronic devices may be located on the top of these raised assets. Typically, these assets are reached by ladders that are permanently anchored to the asset. For example, raised petrochemical/water tanks typically are accessed by ladders that are bolted to one of the legs of the tank. Controlling access to these raised assets by controlling access to the ladder can increase security at a communications sector facility.

A typical ladder access control system consists of some type of cover that is locked or secured over the ladder. The cover can be a casing that surrounds most of the ladder, or a door or shield that covers only part of the ladder. In either case, several rungs of the ladder (the number of rungs depends on the size of the cover) are made inaccessible by the cover, and these rungs can only be accessed by opening or removing the cover. The cover is locked so that only authorized personnel can open or remove it and use the ladder. Ladder access controls are usually installed at several feet above ground level, and they usually extend several feet up the ladder so that they cannot be circumvented by someone accessing the ladder above the control system.

The important features of ladder access control are the size and strength of the cover and its ability to lock or otherwise be secured from unauthorized access.

The covers are constructed from aluminum or some type of steel. This should provide adequate protection from being pierced or cut through. The metals are corrosion resistant so that they will not corrode or become fragile from extreme weather conditions in outdoor applications. The bolts used to install each of these systems are galvanized steel. In addition, the bolts for each cover are installed on the inside of the unit so they cannot be removed from the outside.

Locks

A lock is a type of physical security device that can be used to delay or prevent a door, a gate, a window, a manhole, a filing cabinet drawer, or some other physical feature from being opened, moved, or operated. Locks typically operate by connecting two pieces together—such as by connecting a door to a door jamb or a manhole to its casement. Every lock has two modes—engaged (or "locked"), and disengaged (or "opened"). When a lock is disengaged, the asset on which the lock is installed can be accessed by anyone, but when the lock is engaged, only access to the locked asset.

Before discussing locks and their applicability it is important to discuss key control. Based on our experience, many communications sector facilities (and others) have no idea how many keys for various site/equipment locks have been issued to employees over the years. Many facilities simply issue keys to employees at hiring with no accountability for the keys upon the employee's departure. Needless to say, this is not good security policy. You can have the best-made locks available installed throughout your facilities but if you do not have proper key control, you do not have proper security.

Locks are excellent security features because they have been designed to function in many ways and to work on many different types of assets. Locks can also provide different levels of security depending on how they are designed and implemented. The security provided by a lock is dependent on several factors, including its ability to withstand physical damage (i.e., can it be cut off, broken, or otherwise physically disabled) as well as its requirements for supervision or operation (i.e., combinations may need to be changed frequently so that they are not compromised and the locks remain secure). While there is no single definition of the "security" of a lock, locks are often described as minimum, medium, or maximum security. Minimum security locks are those that can be easily disengaged (or "picked") without the correct key or code, or those that can be disabled easily (such as small padlocks that can be cut with bolt cutters). Higher security locks are more complex and thus are more difficult to pick, or are sturdier and more resistant to physical damage.

Many locks, such as many door locks, only need to be unlocked from one side. For example, most door locks need a key to be unlocked only from the outside. A person opens such devices, called single-cylinder locks, from the inside by pushing a button or by turning a knob or handle. Double-cylinder locks require a key to be locked or unlocked from both sides.

Manhole Intrusion Sensors

Manholes are commonly found in communications sector transmission sites. Manholes are designed to provide access to the underground utilities, meter vaults, petrochemical pumping rooms, and so forth, and therefore they are potential entry points to a system. Because many utilities run under other infrastructure (roads, buildings), manholes also provide potential access points to critical infrastructure as well as petrochemical process assets. In addition, because the portion of the system to which manholes provide entry is primarily located underground, access to a system through a manhole increases the chance that an intruder will not be seen. Therefore, protecting manholes can be a critical component of guarding an entire plant site and a surrounding community.

There are multiple methods for protecting manholes, including preventing unauthorized personnel from physically accessing the manhole, and detecting attempts at unauthorized access to the manhole.

A manhole intrusion sensor is a physical security device designed to detect unauthorized access to the facility through a manhole. Monitoring a manhole that provides access to a chemical plant or processing system can mitigate two distinct types of threats. First, monitoring a manhole may detect access of unauthorized personnel to chemical systems or assets through the manhole. Second, monitoring manholes may also allow the detection of intruders attempting to place explosive or other destructive (WMD) devices into the petrochemical system.

Several different technologies have been used to develop manhole intrusion sensors, including mechanical systems, magnetic systems, and fiber optic and infrared sensors. Some of these intrusion sensors have been specifically designed for manholes, while others consist of standard, off-the-shelf intrusion sensors that have been implemented in a system specifically designed for application in a manhole.

Manhole Locks

A "manhole lock" is a physical security device designed to delay unauthorized access to the communications sector facility or system through a manhole.

Passive Security Barriers

One of the most basic threats facing any facility is from intruders accessing the facility with the intention of causing damage to its assets. These threats may include intruders actually entering the facility, as well as intruders attacking the facility from outside without actually entering it (i.e., detonating a bomb near enough to the facility to cause damage within its boundaries).

Security barriers are one of the most effective ways to counter the threat of intruders accessing a facility or the facility perimeter. Security barriers are large, heavy structures that are used to control access through a perimeter by either vehicles or personnel. They can be used in many different ways depending on how/where they are located at the facility. For example, security barriers can be used on or along driveways or roads to direct traffic to a checkpoint (i.e., a facility may install jersey barriers in a road to direct traffic in a certain direction). Other types of security barriers (crash beams, gates) can be installed at the checkpoint so that guards can regulate which vehicles can access the facility. Finally, other security barriers (i.e., bollards or security planters) can be used along the facility perimeter to establish a protective buffer area between the facility and approaching vehicles. Establishing such a protective buffer can help in mitigating the effects of the type of bomb blast described above, both by potentially absorbing some of the blast, and also by increasing the "stand-off" distance between the blast and the facility (the force of an explosion is reduced as the shock wave travels further from the source, and thus the further the explosion is from the target, the less effective it will be in damaging the target).

Security barriers can be either "active" or "passive." "Active" barriers, which include gates, retractable bollards, wedge barriers, and crash barriers, are readily movable, and thus they are typically used in areas where they must be moved often to allow vehicles to pass—such as in roadways at entrances and exits to a facility. In contrast to active security barriers, "passive" security barriers, which include jersey barriers, bollards, and security planters, are not designed to be moved on a regular basis, and thus they are typically used in areas where access is not required or allowed—such as along building perimeters or in traffic control areas. Passive security barriers are typically large, heavy structures that are usually several feet high, and they are designed so that even heavy-duty vehicles cannot go over or though them. Therefore, they can be placed in a roadway parallel to the flow of traffic so that they direct traffic in a certain direction (such as to a guardhouse, a gate, or some other sort of checkpoint), or perpendicular to traffic such that they prevent a vehicle from using a road or approaching a building or area.

Security for Doorways—Side Hinged Doors

Doorways are the main access points to a facility or to rooms within a building. They are used on the exterior or in the interior of buildings to provide privacy and security for the areas behind them. Different types of doorway security systems may be installed in different doorways depending on the needs or requirements of the buildings or rooms. For example, exterior doorways tend to have heavier doors to withstand the elements and to provide some security to the entrance of the building. Interior doorways in office areas may have lighter doors that may be primarily designed to provide privacy rather than security. Therefore, these doors may be made of glass or lightweight wood. Doorways in industrial areas may have sturdier doors than do other interior doorways and may be designed to provide protection or security for areas behind the doorway. For example, fireproof doors may be installed in chemical storage areas or in other areas where there is a danger of fire.

Because they are the main entries into a facility or a room, doorways are often prime targets for unauthorized entry into a facility or an asset. Therefore, securing doorways may be a major step in providing security at a facility.

A doorway includes four main components:

- The door, which blocks the entrance. The primary threat to the actual door is breaking or piercing through the door. Therefore, the primary security features of doors are their strength and resistance to various physical threats, such as fire or explosions.
- The door frame, which connects the door to the wall. The primary threat to a door frame is that the door can be pried away from the frame. Therefore, the primary security feature of a door frame is its resistance to prying.

- The hinges, which connect the door to the door frame. The primary threat to door hinges is that they can be removed or broken, which will allow intruders to remove the entire door. Therefore, security hinges are designed to be resistant to breaking. They may also be designed to minimize the threat of removal from the door.
- The lock, which connects the door to the door frame. Use of the lock is controlled through various security features, such as keys or combinations so that only authorized personnel can open the lock and go through the door. Locks may also incorporate other security features, such as software or other systems to track overall use of the door or to track individuals using the door.

Each of these components is integral in providing security for a doorway, and upgrading the security of only one of these components while leaving the other components unprotected may not increase the overall security of the doorway. For example, many facilities upgrade door locks as a basic step in increasing the security of a facility. However, if the facilities do not also focus on increasing security for the door hinges or the door frame, the door may remain vulnerable to being removed from its frame, thereby defeating the increased security of the door lock.

The primary attribute for the security of a door is its strength. Many security doors are 4-20 gauge hollow metal doors consisting of steel plates over a hollow cavity reinforced with steel stiffeners to give the door extra stiffness and rigidity. This increases resistance to blunt force used to try to penetrate through the door. The space between the stiffeners may be filled with specialized materials to provide fire, blast, or bullet resistance to the door.

The Windows and Doors Manufacturers Association has developed a series of performance attributes for doors. These include:

- Structural Resistance
- Forced Entry Resistance
- Hinge Style Screw Resistance
- Split Resistance
- Hinge Resistance
- Security Rating
- Fire Resistance

- Bullet Resistance
- Blast Resistance

The first five bullets provide information on a door's resistance to standard physical breaking and prying attacks. These tests are used to evaluate the strength of the door and the resistance of the hinges and the frame in a standardized way. For example, the Rack Load Test simulates a prying attack on a corner of the door. A test panel is restrained at one end, and a third corner is supported. Loads are applied and measured at the fourth corner. The Door Impact Test simulates a battering attack on a door and frame using impacts of 200-foot pounds by a steel pendulum. The door must remain fully operable after the test. It should be noted that door glazing is also rated for resistance to shattering. Manufacturers will be able to provide security ratings for these features of a door as well.

Door frames are an integral part of doorway security because they anchor the door to the wall. Door frames are typically constructed from wood or steel, and they are installed such that they extend for several inches over the doorway that has been cut into the wall. For added security, frames can be designed to have varying degrees of overlap with, or wrapping over, the underlying wall. This can make prying the frame from the wall more difficult. A frame formed from a continuous piece of metal (as opposed to a frame constructed from individual metal pieces) will prevent prying between pieces of the frame.

Many security doors can be retrofitted into existing frames; however, many security door installations include replacing the door frame as well as the door itself. For example, bullet-resistance per Underwriter's Laboratory (UL) 752 requires resistance of the door and frame assembly, and thus replacing the door only would not meet UL 752 requirements.

Valve Lockout Devices

Valves are utilized as control elements in fuel oil/natural gas and petrochemical process piping networks. They regulate the flow of both liquids and gases by opening, closing, or obstructing a flow passageway. Valves are typically located where flow control is necessary. They can be located in-line or at pipeline and tank entrance and exit points. They can serve multiple purposes in a process pipe network, including:

- Redirecting and throttling flow
- Preventing backflow
- Shutting off flow to a pipeline or tank (for isolation purposes)
- Releasing pressure
- Draining extraneous liquid from pipelines or tanks
- Introducing chemicals into the process network
- As access points for sampling process water

Valves may be located either aboveground or below ground. It is critical to provide protection against valve tampering. For example, tampering with a pressure relief valve could result in a pressure buildup and potential explosion in the piping network. On a larger scale, addition of a contaminant or non-compatible chemical substance to the chemical processing system through an unprotected valve could result in the catastrophic release of that contaminant to the general population.

Different security products are available to protect aboveground versus below ground valves. For example, valve lockout devices can be purchased to protect valves and valve controls located aboveground. Vaults containing underground valves can be locked to prevent access to these valves.

As described above, a lockout device can be used as a security measure to prevent unauthorized access to aboveground valves located within petrochemical processing systems. Valve lockout devices are locks that are specially designed to fit over valves and valve handles to control their ability to be turned or seated. These devices can be used to lock the valve into the desired position. Once the valve is locked, it cannot be turned unless the locking device is unlocked or removed by an authorized individual.

Various valve lockout options are available for industrial use, including:

- Cable lockouts
- Padlocked chains/cables
- Valve-specific lockouts

Many of these lockout devices are not specifically designed for use in communications sector industry (i.e., chains, padlocks), but are available from a local hardware store or manufacturer specializing in safety equipment. Other lockout devices (for example, valve-specific lockouts or valve box-locks) are more specialized and must be purchased from safety or valve-related equipment vendors.

The three most common types of valves for which lockout devices are available are gate, ball, and butterfly valves. Each is described in more detail below.

- Gate Valve Lockouts—Gate valve lockouts are designed to fit over the operating hand wheel of the gate valve to prevent it from being turned. The lockout is secured in place with a padlock. Two types of gate valve lockouts are available: diameter-specific and adjustable. Diameter-specific lockouts are available for handles ranging from 1 inch to 13 inches in diameter. Adjustable gate valve lockouts can be adjusted to fit any handle ranging from 1 inch to 6+ inches in diameter.
- Ball Valve Lockouts—There are several different configurations available to lock out ball valves, all of which are designed to prevent rotation of the valve handle. The three major configurations available are a wedge shape for 1-inch to 3-inch valves, a lockout that completely covers ⅜-inch to 8-inch ball valve handles, and a universal lockout that can be applied to quarter-turn valves of varying sizes and geometric handle dimensions. All three types of ball valve lockouts can be installed by sliding the lockout device over the ball valve handle and securing it with a padlock.
- Butterfly Valve Lockouts—The butterfly valve lockout functions in a similar manner to the ball valve lockout. The polypropylene lockout device is placed over the valve handle and secured with a padlock. This type of lockout has been commonly used in the bottling industry.

A major difference between valve-specific lockout devices and the padlocked chain or cable lockouts discussed earlier is that they do not need to be secured to an anchoring device in the floor or the piping system. In addition, valve-specific lockouts eliminate potential tripping or access hazards that may be caused by chains or cable lockouts applied to valves located near walkways or frequently maintained equipment.

Valve-specific lockout devices are available in a variety of colors, which can be useful in distinguishing different valves. For example, different colored lockouts can be used to distinguish the type of liquid passing through the valve (i.e., treated, untreated, potable, petrochemical), or to identify the party responsible for maintaining the lockout. Implementing a system of different-colored locks on operating valves can increase system security by reducing the likelihood of an operator inadvertently opening the wrong valve and causing a problem in the system.

Security For Vents

Vents are installed in some aboveground communications sector storage areas to allow safe venting of off-gases. The specific vent design for any given application will vary depending on the design of the chemical storage vessel. However, every vent consists of an open-air connection between the storage container and the outside environment. Although these air exchange vents are an integral part of covered or underground chemical storage containers, they also represent a potential security threat. Improving vent security by making the vents tamper resistant or by adding other security features, such as security screens or security covers, can enhance the security of the entire petrochemical processing system.

Many municipalities already have specifications for vent security at their local chemical industrial assets. These specifications typically include the following requirements:

- Vent openings are to be angled down or shielded to minimize the entrance of surface and/or rainwater into the vent through the opening
- Vent designs are to include features to exclude insects, birds, animals, and dust
- Corrosion-resistant materials are to be used to construct the vents.

Visual Surveillance Monitoring

Visual surveillance is used to detect threats through continuous observation of important or vulnerable areas of an asset. The observations can also be recorded for later review or use (for example, in court proceedings). Visual surveillance systems can be used to monitor various parts of production, distribution, or pumping/compressing systems, including the perimeter of a facility, outlying pumping stations, or entry or access points into specific buildings. These systems are also useful in recording individuals who enter or leave a facility, thereby helping to identify unauthorized access. Images can be transmitted live to a monitoring station, where they can be monitored in real time, or they can be recorded and reviewed later. Many communications sector facilities have found that a combination of electronic surveillance and security guards provides an effective means of facility security.

Visual surveillance is provided through a closed circuit television (CCTV) system, in which the capture, transmission, and reception of an image is localized within a closed "circuit." This is different than other broadcast images, such as over-the-air television, which is broadcast over the air to any receiver within range.

At a minimum, a CCTV system consists of:

- One or more cameras
- A monitor for viewing the images
- A system for transmitting the images from the camera to the monitor.

Specific attributes and features of camera systems, lenses, and lighting systems are presented in Table 9.11.

Table 9.11. Attributes of Camera, Lenses, and Lighting Systems

Attribute	*Discussion*
	Camera Systems
Camera type	Major factors in choosing the correct camera are the resolution of the image required and lighting of the area to be viewed. • **Solid State** (including charge coupled devices, charge priming device, charge injection device, and metal oxide substrate)—these cameras are becoming predominant in the marketplace because of their high resolution and their elimination of problems inherent in tube cameras. • **Thermal**—These cameras are designed for night vision. They require no light and use differences in temperature between objects in the field of view to produce a video image. Resolution is low compared to other cameras, and the technology is currently expensive relative to other technologies. • **Tube**—These cameras can provide high-resolution burn out and must be replaced after one to two years. In addition, tube performance can degrade over time. Finally, tube cameras are prone to burn images in the tube replacement.
Resolution (the ability to see fine details)	User must determine the amount of resolution required depending on the level of detail required for threat determination. A high definition focus with a wide field of vision will give an optimal viewing area.

(Continued)

Table 9.11. (Continued)

Attribute	*Discussion*
Field of vision width	Cameras are designed to cover a defined field of vision, which is usually defined in degrees. The wider the field of vision, the more area a camera will be able to monitor.
Type of image produced (color, black-and-white, thermal)	Color images may allow the identification of distinctive markings, while black-and-white images may provide sharper contrast. Thermal imaging allows the identification of heat sources (such as human beings or other living creatures) from low-light environments; however, thermal images are not effective in identifying specific individuals (i.e., for subsequent legal processes).
Pan/tilt/zoom (PTZ)	Panning (moving the camera in a horizontal plane), tilting (moving the camera in a vertical plane), and zooming (moving the lens to focus on objects that are at different distances from the camera) allow the camera to follow a moving object. Different systems allow these functions to be controlled manually or automatically. Factors to be considered in PTZ cameras are the degree of coverage for pan and tilt functions and the power of the zoom lens.
	Lenses
Format	Lens format determines the maximum image size to be transmitted.
Focal length	This is the distance from the lens to the center of the focus. The greater the focal length, the higher the magnification, but the narrower the field of vision.
F number	F number is the ability to gather light. Smaller F numbers may be required for outdoor applications where light cannot be controlled as easily.
Distance and width approximation	The distance and width approximations are used to determine the geometry of the space that can be monitored at the best resolution.
	Lighting Systems
Intensity	Light intensity must be great enough for the camera type to produce sharp images. Light can be generated from natural or artificial sources. Artificial sources can be controlled to produce the amount and distribution of light required for a given camera and lens.
Evenness	Light must be distributed evenly over the field of view so that there are no darker or shadowy areas. If there are lighter vs. darker areas, brighter areas may appear washed out (i.e., details cannot be distinguished) while no specific objects can be viewed from darker areas.
Location	Light sources must be located above the camera so that light does not shine directly into the camera.

Source: USEPA (2005).

COMMUNICATION INTEGRATION

In this section, those devices necessary for communication and integration of communications sector industrial processing operations, such as electronic controllers, two-way radios, and wireless data communications are discussed. In regard to security applications, electronic controllers are used to automatically activate equipment (such as lights, surveillance cameras, audible alarms, or locks) when they are triggered. Triggering could be in response to variety of scenarios, including tripping of an alarm or a motion sensor; breaking of a window or a glass door; variation in vibration sensor readings; or simply through input from a timer.

Two-way wireless radios allow two or more users that have their radios tuned to the same frequency to communicate instantaneously with each other without the radios being physically lined together with wires or cables.

Wireless data communications devices are used to enable transmission of data between computer systems and/or between a SCADA server and its sensing devices, without individual components being physically linked together via wires or cables. In industrial petrochemical processing systems, these devices are often used to link remote monitoring stations (i.e., SCADA components) or portable computers (i.e., laptops) to computer networks without using physical wiring connections.

Electronic Controllers

An electronic controller is a piece of electronic equipment that receives incoming electric signals and uses preprogrammed logic to generate electronic output signals based on the incoming signals. While electronic controllers can be implemented for any application that involves inputs and outputs (for example, control of a piece of machinery in a factory), in a security application, these controllers essentially act as the system's "brain," and can respond to specific security-related inputs with a preprogrammed output response. These systems combine the control of electronic circuitry with a logic function such that circuits are opened and closed (and thus equipment is turned on and off) through some preprogrammed logic. The basic principle behind the operation of an electrical controller is that it receives electronic inputs from sensors or any device generating an electrical signal (for example, electrical signals from motion sensors), and then uses preprogrammed logic to produce electrical outputs (for example, these outputs could turn on power to a surveillance camera or to an audible alarm). Thus, these systems automatically generate a preprogrammed, logical response to a preprogrammed input scenario.

The three major types of electronic controllers are timers, electromechanical relays, and programmable logic controllers (PLCs), which are often called "digital relays." Each of these types of controller is discussed in more detail below.

Timers use internal signal/inputs (in contrast to externally generated inputs) and generate electronic output signals at certain times. More specifically, timers control electric current flow to any application to which they are connected, and can turn the current on or off on a schedule pre-specificed by the user. Typical timer range (amount of time that can be programmed to elapse before the timer activates linked equipment) is from 0.2 seconds to 10 hours, although some of the more advanced timers have ranges of up to 60 hours. Timers are useful in fixed applications that don't require frequent schedule changes. For example, a timer can be used to turn on the lights in a room or building at a certain time every day. Timers are usually connected to their own power supply (usually 120–240 V).

In contrast to timers, which have internal triggers based on a regular schedule, electromechanical relays and PLCs have both external inputs and external outputs. However, PLCs are more flexible and more powerful than are electromechanical relays, and thus this section focuses primarily on PLCs as the predominant technology for security-related electronic control applications.

Electromechanical relays are simple devices that use a magnetic field to control a switch. Voltage applied to the relay's input coil creates a magnetic field, which attracts an internal metal switch. This causes the relay's contacts to touch, closing the switch and completing the electrical circuit. This activates any linked equipment. These types of systems are often used for high voltage applications, such as in some automotive and other manufacturing processes.

Two-Way Radios

Two way radios, as discussed here, are limited to a direct unit-to-unit radio communication, either via single unit-to-unit transmission and reception, or via multiple hand-held units to a base station radio contact and distribution system. Radio frequency spectrum limitations apply to all hand-held units, and are directed by the FCC. This also distinguishes a hand-held unit from a base station or base-station unit (such as those used by an amateur (ham) radio operator), which operates under different wave-length parameters.

Two-way radios allow a user to contact another user or group of users instantly on the same frequency, and to transmit voice or data without the need for wires. They use "half-duplex" communications, or communication that can be only transmitted or received; they cannot transmit and receive simultaneously. In other words, only one person may talk, while other personnel with radio(s) can only listen. To talk, the user depresses the talk button and speaks into the radio. The audio then transmits the voice wirelessly to the receiving radios. When the speaker has finished speaking and the channel has cleared, users on any of the receiving radios can transmit; either to answer the first transmission, or to begin a new conversation. In addition to carrying voice data, many types of wireless radios also allow the transmission of digital data, and these radios may be interfaced with computer networks that can use or track these data. For example, some two-way radios can send information such as global positioning system (GPS) data, or the ID of the radio. Some two-way radios can also send data through a SCADA system.

Wireless radios broadcast these voice or data communications over the airwaves from the transmitter to the receiver. While this can be an advantage in that the signal emanates in all directions and does not need a direct physical connection to be received at the receiver, it can also make the communications vulnerable to being blocked, intercepted, or otherwise altered. However, security features are available to ensure that the communications are not tampered with.

Wireless Data Communications

A wireless data communication system consists of two components: a "Wireless Access Point" (WAP), and a "Wireless Network Interface Card" (sometimes also referred to as a "Client"), which work together to complete the communications link. These wireless systems can link electronic devices, computers, and computer systems together using radio waves, thus eliminating the need for these individual components to be directly connected together through physical wires. While wireless data communications have widespread application in communication systems, they also have limitations. First, wireless data connections are limited by the distance between components (radio waves scatter over a long distance and cannot be received efficiently, unless special directional antennas are used). Second, these devices only function if the individual components are in a direct line of sight with each other, since radio waves are affected by interference from physical obstructions. However, in some cases, repeater units can be used to amplify and retransmit wireless signals to circumvent these problems. The two components of wireless devices are discussed in more detail below.

(1) WAP: The WAP provides the wireless data communication service. It usually consists of a housing (which is constructed from plastic or metal depending on the environment it will be used in) containing a circuit board; flash memory that holds software; one of two external ports to connect to existing wired networks; a wireless radio transmitter/receiver; and one or more antenna connections. Typically, the WAP requires a one-time user configuration to allow the device to interact with the Local Area Network (LAN). This configuration is usually done via a web-driven software application which is accessed via a computer.
(2) Wireless Network Interface Card/Client: A wireless card is a piece of hardware that is plugged into a computer and enables that computer to make a wireless network connection. The card consists of a transmitter, functional circuitry, and a receiver for the wireless signal, all of which work together to enable communication between the computer, its wireless transmitter/receiver, and its antenna connection. Wireless cards are installed in a computer through a variety of connections, including USB Adapters, or Laptop CardBus (PCMCIA) or Desktop Peripheral (PCI) cards. As with the WAP, software is loaded onto the user's computer, allowing configuration of the card so that it may operate over the wireless network.

Two of the primary applications for wireless data communications systems are to enable mobile or remote connections to a LAN, and to establish wireless communications links between SCADA remote telemetry units (RTUs) and sensors in the field. Wireless car connections are usually used for LAN access from mobile computers. Wireless cards can also be incorporated into RTUs to allow them to communicate with sensing devices that are located remotely.

CYBER PROTECTION DEVICES

Various cyber protection devices are currently available for use in protecting communications sector computer systems. These protection devices include anti-virus and pest eradication software, firewalls, and network intrusion hardware/software. These products are discussed in this section.

Anti-Virus and Pest Eradication Software

Anti-virus programs are designed to detect, delay, and respond to programs or pieces of code that are specifically designed to harm computers. These programs are known as "malware." Malware can include computer viruses, worms, and Trojan Horse programs (programs that appear to be benign but which have hidden harmful effects).

Pest eradication tools are designed to detect, delay, and respond to "spyware" (strategies that websites use to track user behavior, such as by sending "cookies" to the user's computer), and hacker tools that track keystrokes (keystroke loggers) or passwords (password crackers).

Viruses and pests can enter a computer system through the Internet or through infected floppy discs or CDs. They can also be placed onto a system by insiders. Some of these programs, such as viruses and worms, then move within a computer's drives and files or between computers if the computers are networked to each other. This malware can deliberately damage files, utilize memory and network capacity, crash application programs, and initiate transmissions of sensitive information from a PC. While the specific mechanisms of these programs differ, they can infect files, and even the basic operating program of the computer firmware/hardware.

The most important features of an anti-virus program are its abilities to identify potential malware and to alert a user before infection occurs, as well as its ability to respond to a virus already resident in a system. Most of these programs provide a log so that the user can see what viruses have been detected and where they were detected. After detecting a virus, the anti-virus software may delete the virus automatically, or it may prompt the user to delete the virus. Some programs will also fix files or programs damaged by the virus.

Various sources of information are available to inform the general public and computer system operators about new viruses being detected. Since anti-virus programs use signatures (or snippets of code or data) to detect the presence of a virus, periodic updates are required to identify new threats. Many anti-virus software providers offer free upgrades that are able to detect and respond to the latest viruses.

Firewalls

A firewall is an electronic barrier designed to keep computer hackers, intruders or insiders from accessing specific data files and information on a communication sector's computer network or other electronic/computer systems. Firewalls operate by evaluating and then filtering information coming through a public network (such as the Internet) into the utility's computer or other electronic system. This evaluation can include identifying the source or destination addresses and ports, and allowing or denying access based on this identification.

There are two methods used by firewalls to limit access to the utility's computers or other electronic systems from the public network:

- The firewall may deny all traffic unless it meets certain criteria
- The firewall may allow all traffic through unless it meets certain criteria.

A simple example of the first method is to screen requests to ensure that they come from an acceptable (i.e., previously identified) domain name and Internet protocol address. Firewalls may also use more complex rules that analyze the application data to determine if the traffic should be allowed through. For example, the firewall may require user authentication (i.e., use of a password) to access the system. How a firewall determines what traffic to let through depends on which network layer it operates at and how it is configured. Some of the pros and cons of various methods to control traffic flowing in and out of the network are provided in Table 9.12.

Firewalls may be a piece of hardware, a software program, or an appliance card that contains both.

Advanced features that can be incorporated into firewalls allow for the tracking of attempts to log on to the local area network system. For example, a report of successful and unsuccessful log-in attempts may be generated for the computer specialist to analyze. For systems with mobile users, firewalls allow remote access into the private network by the use of secure log-on procedures and authentication certificates. Most firewalls have a graphical user interface for managing the firewall.

Table 9.12. Pros and Con of Various Firewall Methods for Controlling Network Access

Method	*Description*	*Pros*	*Cons*
Packet filtering	Incoming and outgoing packets (small chunks of data) are analyzed against a set of filters. Packets that make it through the filters are sent to the requesting system and all others are discarded. There are two types of packet filtering: static (the most common) and dynamic.	Static filtering is relatively inexpensive, and little maintenance required. It is well-suited for closed environments where access to or from multiple addresses is not allowed.	Leaves permanent open holes in the network; allows direct connection to internal hosts by external sources; offers no user authentication; method can be unmanageable in large networks.
Proxy service	Information from the Internet is retrieved by the firewall and then sent to the requesting system and vice versa. In this way, the firewall can limit the information made known to the requesting system, making vulnerabilities less apparent.	Only allows temporary open holes in the network perimeter. Can be used for all types of internal protocol services.	Allows direct connections to internal hosts by external clients; offers no user authentication.
Stateful pattern recognition	This method examines and compares the contents of certain key parts of an information packet against a database of acceptable information. Information traveling from inside the firewall to the outside is monitored for specific defining characteristics, then incoming information is compared to these characteristics. If the comparison yields a reasonable match, the information is allowed through. If not, the information is discarded.	Provides a limited time window to allow pockets of information to be sent; does not allow any direct connections between internal and external hosts; supports user-level authentication.	Slower than packet filtering; does not support all types of connections.

USEPA (2005).

In addition, new Ethernet firewall cards that fit in the slot of an individual computer bundle additional layers of defense (like encryption and permit/deny) for individual computer transmissions to the network interface function. These new cards have only a slightly higher cost than traditional network interface cards.

Network Intrusion Hardware/Software

Network intrusion detection and prevention systems are software- and hardware-based programs designed to detect unauthorized attacks on a computer network system.

While other applications, such as firewalls and anti-virus software, share similar objectives with network intrusion systems, network intrusion systems provide a deeper layer of protection beyond the capabilities of these other systems because they evaluate patterns of computer activity rather than specific files.

It is worth noting that attacks may come from either outside or within the system (i.e., from an insider), and that network intrusion detection systems may be more applicable for detecting patterns of suspicious activity from inside a facility (i.e., accessing sensitive data, etc.) than are other information technology solutions.

Network intrusion detection systems employ a variety of mechanisms to evaluate potential threats. The types of search and detection mechanisms are dependent upon the level of sophistication of the system. Some of the available detection methods include:

- Protocol analysis—Protocol analysis is the process of capturing, decoding, and interpreting electronic traffic. The protocol analysis method of network intrusion detection involves the analysis of data captured during transactions between two or more systems or devices, and the evaluation of these data to identify unusual activity and potential problems. Once a problem is isolated and recorded, problems or potential threats can be linked to pieces of hardware or software. Sophisticated protocol analysis will also provide statistics and trend information on the captured traffic.
- Traffic anomaly detection—Traffic anomaly detection identifies potential threatening activity by comparing incoming traffic to "normal" traffic patterns, and identifying deviations. It does this by comparing user character-

istics against thresholds and triggers defined by the network administrator. This method is designed to detect attacks that span a number of connections, rather than a single session.

- Network honeypot—This method establishes non-existent services in order to identify potential hackers. A network honeypot impersonates services that don't exist by sending fake information to people scanning the network. It identifies the attacker when they attempt to connect to the service. There is no reason for legitimate traffic to access these resources because they don't exist; therefore any attempt to access them constitutes an attack.
- Anti-intrusion detection system evasion techniques—These methods are designed to find attackers who may be trying to evade intrusion detection system scanning. They include methods called IP defragmentation, TCP streams reassembly, and deobfuscation.

While these detection systems are automated, they can only indicate patterns of activity and a computer administrator or other experienced individual must interpret activities to determine whether or not they are potentially harmful. Monitoring the logs generated by these systems can be time consuming, and there may be a learning curve to determine a baseline of "normal" traffic patterns from which to distinguish potential suspicious activity.

REFERENCES AND RECOMMENDED READING

DHS. (2010). Communications Sector-Specific Plan. Washington, DC: U.S. Department of Homeland Security.

Garcia, M. L. (2001). *The Design and Evaluation of Physical Protection Systems.* Oxford: Butterworth-Heinemann.

IBWA. (2004). *Bottled Water Safety and Security.* Alexandria, VA: International Bottled Water Association.

NAERC. (2002). *Security Guidelines for the Electricity Sector.* Washington, DC: North American Electric Reliability Council.

Schneier, B. (2000). *Secrets & Lies.* New York: Wiley.

USEPA. (2005). *Water and Wastewater Security Product Guide.* Accessed 4/14/16. http://cfpub.epa.gov.safewater/watersecurity/guide.

10

The Paradigm Shift

It takes disasters to trigger change because dangers that remain hypothetical fail to trigger appropriate sensory response.

—D. D. P. Johnson and E. M. P. Madin

The 9/11 shift: There is a new world view in the making.

If men would learn from history, what lessons it might teach us!

—Samuel Coleridge

The events of 9/11 dramatically changed this nation and focused us on combating terrorism. As a result, in 2003 and subsequent years, the Department of Homeland Security (DHS) in conjunction with members from the general public, state and local agencies, and private groups concerned with the safety of critical infrastructures, established a Water Security Working Group (WSWG) to consider and make recommendations on infrastructure security issues. Although initially created to make recommendations for water/wastewater security, WSWG is an excellent template for use with other critical infrastructures, including communication assets. For example, the WSWG identified active and effective security practices for critical infrastructure, and provided an approach for adopting these practices. It also recommended mechanisms to provide incentives that facilitate broad and receptive response among critical infrastructure sectors to implement active and effective security practices. Finally, WSWG recommended mechanisms to measure progress and achievements in implementing active and effective security practices, and identifying barriers to implementation.

The WSWG recommendations on security are structured to maximize benefits to critical industries by emphasizing actions that have the potential both to improve the quality or reliability of service, and to enhance security. These recommendations, based on original recommendations from the 2003 National Drinking Water Advisory Council (NDWAC), were designed primarily, as the name suggests, for use by water systems of all types and sizes, including systems that serve less than 3,300 people. However, it is the author's opinion, based on personal experience, that NDWAC's recommendations, when properly adapted to applicable circumstances and locations, can be applied to any and all critical infrastructure sectors, including the communications sector.

The NDWAC identified fourteen features of active and effective security programs that are important to increasing security and relevant across the broad range of utility circumstances and operating conditions. USEPA (2003) points out that the fourteen features are, in many cases, consistent with the steps needed to maintain technical, management, and operational performance capacity related to overall water quality; as pointed out earlier, these steps can be applied to other critical infrastructures as well. Many facilities may be able to adopt some of the features with minimal, if any, capital investment.

FOURTEEN FEATURES OF ACTIVE AND EFFECTIVE SECURITY

It is important to point out that the fourteen features of active and effective programs emphasize that "one size does not fit all" and are not a cookie cutter approach to effective implementation of security measures. There will be variability in security approaches and tactics among communications sector facilities, based on industry-specific circumstances and operating conditions. The fourteen features:

- Are sufficiently flexible to apply to all communication assets, regardless of size.
- Incorporate the idea that active and effective security programs should have measurable goals and time lines.

- Allow flexibility for communications sector industrial facilities to develop specific security approaches and tactics that are appropriate to industry-specific circumstances.

Communications sector facilities can differ in many ways, including:

- Number of supply sources
- Energy capacity
- Operation risk
- Location risk
- Security budget
- Spending priorities
- Political and public support
- Legal barriers
- Public versus private ownership

Communications sector facilities should address security in an informed and systematic way, regardless of these differences. Communications sector facilities need to fully understand the specific, local circumstances and conditions under which they operate, and develop a security program tailored to those conditions. The goal in identifying common features of active and effective security programs is to achieve consistency in security program outcomes among communications sector facilities, while allowing for and encouraging facilities to develop utility-specific security approaches and tactics. The features are based on a comprehensive "security management layering system" approach that incorporates a combination of public involvement and awareness, partnerships, and physical, chemical, operational, and design controls to increase overall program performance. They address industry security in four functional categories: *organization, operation, infrastructure,* and *external.* These functional categories are discussed in greater detail below.

- **Organizational**—There is always something that can be done to improve security. Even when resources are limited, the simple act of increasing organizational attentiveness to security may reduce vulnerability and increase responsiveness. Preparedness itself can help deter attacks. The first step to achieving preparedness is to make security a part of the organizational culture, so that it is in the day-to-day thinking of front-line employees, emergency responders, and management of every communication sector facility in this country. To successfully incorporate security into "business as usual," there must be a strong commitment to security by organization leadership and by the supervising body, such as the board of stockholders. The following features address how a security culture can be incorporated into an organization.

- **Operational**—In addition to having a strong culture and awareness of security within an organization, an active and effective security program makes security part of operational activities, from daily operations, such as monitoring physical access controls to scheduled annual reassessments. Communication sector entities will often find that by implementing security into operations they can also reap cost benefits and improve the quality or reliability of the energy service.
- **Infrastructure**—These recommendations advise utilities to address security in all elements of communications sector infrastructure—from source to distribution and through processing and product delivery.
- **External**—Strong relationships with response partners and the public strengthen security and public confidence. Two of the recommended features of active end effective security programs address this need.

Fourteen Features

Feature 1. Make an explicit and visible commitment of the senior leadership to security.

Communication sector facilities should create an explicit, easily communicated, enterprise-wide commitment to security, which can be done through:

- Incorporating security into a utility-wide mission or vision statement, addressing the full scope of an active and effective security program—that is, protection of worker/public health, worker/public safety, and public confidence, and that is part of core day-to-day operations.
- Developing an enterprise-wide security policy, or set of policies.

Communications sector entities should use the process of making a commitment to security as an opportunity to raise awareness of security throughout the organization, making the commitment visible to all employees and customers, and to help every facet of the enterprise to recognize the contribution they can make to enhancing security.

Feature 2. Promote security awareness throughout the organization. The objective of a security culture should be to make security awareness a normal, accepted, and routine part of day-to-day operations. Examples of tangible efforts include:

- Conducting employee training
- Incorporating security into job descriptions
- Establishing performance standards and evaluations for security
- Creating and maintaining a security tip line and suggestion box for employees
- Making security a routine part of staff meetings and organization planning
- Creating a security policy.

Feature 3. Assess vulnerabilities and periodically review and update vulnerability assessments to reflect changes in potential threats and vulnerabilities.
Because circumstances change, communications sector facilities should maintain their understanding and assessment of vulnerabilities as a "living document," and continually adjust their security enhancement and maintenance priorities. Communications sector facilities should consider their individual circumstances and establish and implement a schedule for review of their vulnerabilities.

Assessments should take place once every three to five years at a minimum. Communications sector facilities may be well served by doing assessments annually.

The basic elements of sound vulnerability assessments are:

- Characterization of the chemical processing system, including its mission and objectives
- Identification and prioritization of adverse consequences to avoid
- Determination of critical assets that might be subject to malevolent acts that could result in undesired consequences

- Assessment of the likelihood (qualitative probability) of such malevolent acts from adversaries
- Evaluation of existing countermeasures
- Analysis of current risk and development of a prioritized plan for risk reduction.

Feature 4. Identify security priorities and, on an annual basis, identify the resources dedicated to security programs and planned security improvements, if any.
Dedicated resources are important to ensure a sustained focus on security. Investment in security should be reasonable, considering utilities' specific circumstances. In some circumstances, investment may be as simple as increasing the amount of time and attention that executives and managers give to security. Where threat potential or potential consequences are greater, greater investment likely is warranted.

This feature establishes the expectation that communication sector facilities should, through their annual capital, operations, maintenance, and staff resources plans, identify and set aside resources consistent with their specific identified security needs. Security priorities should be clearly documented and should be reviewed with utility executives at least once per year as part of the traditional budgeting process.

Feature 5. Identify managers and employees who are responsible for security and establish security expectations for all staff.

- Explicit identification of security responsibilities is important for development of a security culture with accountability
- At minimum, communication sector facilities should identify a single, designated individual responsible for overall security, even if other security roles and responsibilities will likely be dispersed throughout the organization
- The number and depth of security-related roles will depend on a utility's specific circumstances.

Feature 6. Establish physical and procedural controls to restrict access to chemical industrial infrastructure to only those conducting authorized, official business and to detect unauthorized physical intrusions.

Examples of physical access controls include fencing critical areas, locking gates and doors, and installing barriers at site access points. Monitoring for physical intrusion can include maintaining well-lighted facility perimeters, installing motion detectors, and utilizing intrusion alarms. The use of neighborhood watches, regular employee rounds, and arrangements with local police and fire departments can support identifying unusual activity in the vicinity of facilities.

Examples of procedural access controls include inventorying keys, changing access codes regularly, and requiring security passes to pass gates and access sensitive area. In addition, utilities should establish the means to readily identify all employees including contractors and temporary workers with unescorted access to facilities.

Feature 7. Employee protocols for detection of contamination consistent with the recognized limitations in current contaminant detection, monitoring, and surveillance technology.

Until progress can be made in development of practical and affordable online contaminant monitoring and surveillance systems, most communications sector facilities must use other approaches to contaminant monitoring and surveillance.

Many utilities already measure the above parameters (and many others) on a regular basis to control plant operations and confirm chemical mixture quality; more closely monitoring these parameters may create operational benefits for facilities that extend far beyond security, such as reducing operating costs and chemical usage. Communications sector facilities also should thoughtfully monitor customer complaints and improve connections with local public health networks to detect public health anomalies. Customer complaints and public health anomalies are an important way to detect potential contamination problems and other environmental quality concerns.

Feature 8. Define security-sensitive information; establish physical, electronic, and procedural controls to restrict access to security-sensitive information; detect unauthorized access; and ensure information and communications systems will function during emergency response and recover.

Protecting IT systems largely involves using physical hardening and procedural steps to limit the number of individuals with authorized access and to prevent access by unauthorized individuals. Examples of physical steps to harden SCADA and IT networks include installing and maintaining fire walls, and screening the network for viruses. Examples of procedural steps include restricting remote access to data networks, and safeguarding critical data through backups and storage in safe places. Utilities should strive for continuous operation of IT and telecommunications systems, even in the event of an attack, by providing an uninterruptible power supply and back-up systems, such as satellite phones.

In addition to protecting IT systems, security sensitive information should be identified and restricted to the appropriate personnel. Security-sensitive information could be contained within:

- Facility maps and blueprints
- Operations details
- Hazardous material utilization and storage
- Tactical level security program details
- Any other information on utility operations or technical details that could aid in planning or execution of an attack.

Identification of security-sensitive information should consider all ways that utilities might use and make public information (e.g., many chemical industrial facilities may at times engage in competitive bidding processes for construction of new facilities or infrastructure). Finally, information critical to the continuity of day-to-day operations should be identified and backed up.

Feature 9. Incorporate security considerations into decisions about acquisition, repair, major maintenance, and replacement of physical infrastructure; include consideration of opportunities to reduce risk through physical hardening and adoption of inherently lower-risk design and technology options.

Prevention is a key aspect of enhancing security. Consequently, consideration of security issues should begin as early as possible in facility construction (i.e., it should be a factor in building plans and designs). However, to incorporate security considerations into design choices, communication facilities need information about the types of security design approaches and equipment that are available and the performance of these designs and equipment in multiple dimensions. For example, communications sector facilities would want to evaluate not just the way that a particular design might contribute to security, but would also look at how that design would affect the efficiency of day-to-day plant operations and worker safety.

Feature 10. Monitor available threat-level information and escalate security procedures in response to relevant threats.

Monitoring threat information should be a regular part of a security program manager's job, and utility-, facility-, and region-specific threat levels and information should be shared with those responsible for security. As part of security planning, communications sector facilities should develop systems to access threat information and procedures that will be followed in the event of increased industry or facility threat levels, and should be prepared to put these procedures in place immediately, so that adjustments are seamless. Involving local law enforcement and FBI is critical.

Communications sector facilities should investigate what networks and information sources might be available to them locally, and at the state and regional level. If a utility cannot gain access to some information networks, attempts should be made to align with those who can and will provide effective information to the communication sector facility.

Feature 11. Incorporate security considerations into emergency response and recovery plans, test and review plans regularly, and update plans to reflect changes in potential threats, physical infrastructure, chemical processing operations, critical interdependencies, and response protocols in partner organizations.

Communications sector facilities should maintain response and recovery plans as "living documents." In incorporating security considerations into their emergency response and recovery plans, chemical facilities also should be aware of the National Incident Management System (MIMS) guidelines, established by DHS, and of regional and local incident management commands and systems, which tend to flow from the national guidelines.

Communications sector facilities should consider their individual circumstances and establish, develop, and implement a schedule for review of emergency response and recovery plans. Communications sector facility plans should be thoroughly coordinated with emergency response and recovery planning in the larger community. As part of this coördination, a mutual aid program should be established to arrange in advance for exchanging resources (personnel or physical assets) among agencies within a region, in the event of an emergency or disaster that disrupts operations. Typically, the exchange of resources is based on a written, formal mutual agreement. For example, Florida's Water-Wastewater Agency Response Network (FlaWARN), deployed after Hurricane Katrina, allowed the new "utilities helping utilities" network to respond to urgent requests from Mississippi for help to bring facilities back on line after the hurricane.

The emergency response and recovery plans should be reviewed and updated as needed annually. This feature also establishes the expectation that chemical facilities should test or exercise their emergency response and recovery plans regularly.

Feature 12. Develop and implement strategies for regular, ongoing security-related communications with employees, response organizations, rate setting organizations, and customers.

An active and effective security program should address protection of public health, public safety (including infrastructure), and public confidence. Communications sector facilities should create an awareness of security and an understanding of the rationale for their overall security management approach in the communities they reside in and/or serve.

Effective communication strategies consider key messages—who is best equipped/trusted to deliver the key messages; the need for message consistency, particularly during an emergency; and the best mechanisms for delivering messages and for receiving information and feedback from key partners. The key audiences for communication strategies are utility employees, response organizations, and customers.

Feature 13. Forge reliable and collaborative partnerships with the communities served, managers of critical interdependent infrastructure, response organizations, and other local utilities.

Effective partnerships build collaborative working relationships and clearly define roles and responsibilities, so that people can work together seamlessly if an emergency should occur. It is important for communications sector facilities within a region and neighboring regions to collaborate and establish a mutual aid program with neighboring utilities, response organizations, and sectors, such as the power sector, on which utilities rely or impact. Mutual aid agreements provide for help from other organizations that is prearranged and can be accessed quickly and efficiently in the event of a terrorist attack or natural disaster. Developing reliable and collaborative partnerships involves reaching out to managers and key staff and other organizations to build reciprocal understanding and to share information about the facility's security concerns and planning. Such efforts will maximize the efficiency and effectiveness of a mutual aid program during an emergency response effort, as the organizations will be familiar with each others' circumstances, and thus will be better able to serve each other.

It is also important for communications sector facilities to develop partnerships with the communities and customers they serve. Partnerships help to build credibility within communities and establish public confidence in utility operations. People who live near communications sector facility structures can be the eyes and ears of the facility, and can be encouraged to notice and report changes in operating procedures or other suspicious behaviors.

Communications sector facilities and public health organizations should establish formal agreements on coordination to ensure regular exchange of information between facilities and public health organizations, and outline roles and responsibilities during response to and recovery from an emergency. Coordination is important at all levels of the public health community—national public health, county health agencies, and healthcare providers, such as hospitals.

Feature 14. Develop chemical facility-specific measures of security activities and achievements, and self-assess against these measures to understand and document program progress.

Although security approaches and tactics will be different depending on chemical utility-specific circumstances and operating conditions, we recommend that all communications sector facilities monitor and measure a number of common types of activities and achievements, including existence of program policies and procedures, training, testing, and implementing schedules and plans.

The Fourteen-Feature Matrix

In the following, a matrix of recommended measures to assess effectiveness of a communications sector facility's security program is presented. Each feature is grouped according to its functional category: organization, operation, infrastructure, and external.

Table 10.1. Fourteen Features of an Active and Effective Security Matrix

Features	*Checklist: Potential Measures of Progress*
Organizational Features	
Feature 1—Explicit commitment to security	Does a written, enterprise-wide security policy exist, and is the policy reviewed regularly and updated as needed?
Feature 2—Promote security awareness	Are incidents reported in a timely way, and are lessons learned from incident responses reviewed and, as appropriate, incorporated into future utility security efforts?
Feature 5—Defined security roles and employee expectations	Are managers and employees who are responsible for security identified?
Operational Features	
Feature 3—Vulnerability assessment up to date	Are reassessments of vulnerabilities made after incidents, and are lessons learned and other relevant information incorporated into security practices?
Feature 4—Security resources and implementation priorities	Are security priorities clearly identified, and to what extent do security priorities have resources assigned to them?
Feature 7—Contamination detection	Is there a protocol/procedure in place to identify and respond to suspected contamination events?
Feature 10—Threat-level based protocols	Is there a protocol/procedure of responses that will be made if threat levels change?
Feature 11—Emergency response plan tested and up to date	Do exercises address the full range of threats—physical, cyber, and contamination—and is there a protocol/procedure to incorporate lessons learned from exercises and actual responses into updates to emergency response and recovery plans?
Feature 14—Industry-specific measures and self-assessment	Does the utility perform self-assessment at least annually?

Features	*Checklist: Potential Measures of Progress*
	Infrastructure Features
Feature 6—Intrusion detection and access control	To what extent are methods to control access to sensitive assets in place?
Feature 8—Information protection and continuity	Is there a procedure to identify and control security-sensitive information, is information correctly categorized, and how to control measures perform under testing?
Feature 9—Design and construction standards	Are security considerations incorporated into internal utility design and construction standards for new facilities/ infrastructures and major maintenance projects?
	External Features
Feature 12—Communications	Is there a mechanism for utility employees, partners and the community to notify the utility of suspicious occurrences and other security concerns?
Feature 13—Partnerships	Have reliable and collaborative partnerships with customers, managers of independent interrelated infrastructure, and response organizations been established?

Source: USEPA (2003).

Ultimately, the goal of implementing the fourteen security features (and all other security provisions) is to create a significant improvement in communications sector facilities on a national scale, by reducing vulnerabilities, and therefore risk to public health from terrorist attacks and natural disasters. To create a sustainable effect, the communications sector as a whole must not only adopt and actively practice the features, but also incorporate the features into "business as usual."

REFERENCES AND RECOMMENDED READING

USEPA. (2003). *Active and Effective Water Security Programs*. Accessed 6/06. http://cfpub.epa.gov/safewater/watersecurity/14 features.cfm.

Glossary

acid bomb—A crude bomb made by combining muriatic acid with aluminum strips in a 2-liter soda bottle.

aerosol—A fine mist or spray, which contains minute particles.

Afghanistan—At the time of 9/11, Afghanistan was governed by the Taliban and Osama bin Laden called it home. Amid U.S. air strikes, which began on October 7, 2001, the U.S. sent in more than $300 million in humanitarian aid. In December, 2001, Afghanistan reopened their embassy for the first time in more than 20 years.

aflatoxin—A toxin created by bacteria that grow on stored foods, especially on rice, peanuts, and cotton seeds.

agency—A division of government with a specific function, or a non-governmental organization (e.g., private contractor, business, etc.) that offers a particular kind of assistance. In the incident command system, agencies are defined as jurisdictional (having statutory responsibility for incident mitigation) or assisting and/or cooperating (providing resources and/or assistance).

air marshal—A federal marshal whose purpose is to ride commercial flights dressed in plain clothes and armed to prevent hijackings. Israel's use of air marshals on El Al is credited as the reason Israel has had a single hijacking in 31 years. The U.S. started using air marshals after September 11. Despite President Bush's urging there are not enough air marshals to go around, so many flights do not have them.

airborne—Carried by or through the air.

al Jazeera—Satellite television station based in Qatar and broadcast throughout the Middle East, al Jazeera has often been called the "CNN" of the Arab world.

al-Qaeda—Meaning "the Base"; an international terrorist group founded in approximately 1989 and dedicated to opposing non-Islamic governments with force and violence. One of the principal goals of al-Qaeda was to drive the U.S armed forces out of the Saudi Arabian peninsula and Somalia by violence. Currently wanted for several terrorist attacks, including those on the U.S. embassy in Kenya and Tanzania as well as the first and second World Trade Center bombings, and the attack on the Pentagon.

al Tahwid—A Palestinian group based in London which professes a desire to destroy both Israel and the Jewish people throughout Europe. Eleven al Tahwid members were arrested in Germany allegedly as they were about to begin attacking that country.

all-hazards—A grouping classification encompassing all conditions, environmental or manmade, that have the potential to cause injury, illness, or death; damage to or loss of equipment, infrastructure services, or property; or alternatively causing functional degradation to social, economic, or environmental aspects.

alpha radiation—The least penetrating type of nuclear radiation. Not considered dangerous unless particles enter the body.

American Airlines Flight 11—The Boeing 767 carrying 81 passengers, 9 flight attendants, and 2 pilots, which was high-jacked and crashed into the north tower of the World Trade Center at 8:45 a.m. Eastern Time on September 11, 2001. Flight 11 was en route to Los Angeles from Boston.

American Airlines Flight 77—The Boeing 757 carrying 58 passengers, 4 flight attendants, and 2 pilots, which was high-jacked and crashed into the Pentagon at 9:40 a.m. Eastern Time on September 11, 2001. Flight was en route to Los Angeles from Dulles International Airport in Virginia.

ammonium nitrate-fuel oil (ANFO)—A powerful explosive made by mixing fertilizer and fuel oil. The type of bomb used in the first World Trade Center attack as well as the Oklahoma City bombing.

analyte—The name assigned to a substance or feature that describes it in terms of its molecular composition, taxonomic nomenclature, or other characteristic.

anthrax—An often fatal infectious disease contracted from animals. Anthrax spores have a long survival period, the incubation period is short, and disability is severe, making anthrax a bioweapon of choice by several nations.

antidote—A remedy to counteract the effects of poison.

antigen—A substance which stimulates an immune response by the body. The immune system recognizes such substances as foreign and produces antibodies to fight them.

antitoxin—An antibody which neutralizes a biological toxin.

Armed Islamic Group (GIA)—An Algerian Islamic extremist group which aims to overthrow the secular regime in Algeria and replace it with an Islamic state. The GIA began its violent activities in early 1992 after Algiers voided the victory of the largest Islamic party, Islamic Salvation Front (FIS), in the December 1991 elections.

asset—A person, structure, facility, information, material or process that has value. In the context of the National Infrastructure Protection Plan (NIPP), people are not considered assets.

asymmetric threat—The use of crude or low-tech methods to attack a superior or more high-tech enemy.

Axis of Evil—Iran, Iraq, and North Korea as mentioned by President G.W. Bush during his State of the Union speech in 2002 as nations which were a threat to U.S. security due to harboring terrorism.

Baath Party—The official political party in Iraq until the U.S. debaathified Iraq in May 2003, after a war which lasted a little over a month. Saddam Hussein, the former ruler of the Baath party, was targeted by American-led coalition forces and fled. Baath party members have been officially banned from participating in any new government in Iraq.

Beltway Sniper—For nearly a month in October, 2002, the Washington, DC, Maryland, and Virginia area was the hunting grounds for 41-year old John Allen Muhammad and 17-year old Lee Boyd Malvo. Dubbed "the Beltway Sniper" by the media, they shot people at seemingly random places such as schools, restaurants, and gas stations.

Bioaccumulative—Substances that concentrate in living organisms as they breathe contaminated air, drink or live in contaminated water, or eat contaminated food rather than being eliminated through natural processes.

biochemical warfare—Collective term for use of both chemical warfare and biological warfare weapons.

biochemterroism—Terrorism using as weapons biological or chemical agents.

biological ammunition—Ammunition designed specifically to release a biological agent used as the warhead for biological weapons. Biological ammunition may take many forms, such as a missile warhead or bomb.

biological attacks—The deliberate release of germs or other biological substances that cause illness.

Biosafety Level 1—Suitable for work involving well-characterized biological agents not known to consistently cause disease in healthy adult humans, and of minimal potential hazard to lab personnel and the environment. Work is generally conducted on open bench tops using standard microbiological practices.

Biosafety Level 2—Suitable for work involving biological agents of moderate potential hazard to personnel and the environment. Lab personnel should have specific training in handling pathogenic agents and be directed by competent scientists. Access to the lab should be limited when work is being conducted, extreme precautions should be taken with contaminated sharp items, and certain procedures should be conducted in biological safety cabinets or other physical containment equipment if there is a risk of creating infectious aerosols or splashes.

Biosafety Level 3—Suitable for work done with indigenous or exotic biological agents that may cause serious or potentially lethal disease as a result of exposure by inhalation. Lab personnel must have specific training in handling pathogenic and potentially lethal agents and be supervised by competent scientists who are experienced in working with these agents. All procedures involving the manipulation of infectious material are conducted within biological safety cabinets or other physical containment devices, or by personnel wearing appropriate personal protective clothing and equipment. The lab must have special engineering and design features.

Biosafety Level 4—Suitable for work with the most infectious biological agents. Access to the two Biosafety Level 4 labs in the U.S. is highly restricted.

bioterrorism—The use of biological agents in a terrorist operation. Biological toxins would include anthrax, Ricin, botulism, the plague, smallpox, and tularemia.

Bioterrorism Act—The Public Health Security and Bioterrorism Preparedness and Response Act of 2002.

biowarfare—The use of biological agents to cause harm to targeted people either directly, by bringing the people into contact with the agents or indirectly, by infecting other animals and plants, which would in turn cause harm to the people.

blister agents—Agents which cause pain and incapacitation instead of death and might be used to injure many people at once, thereby overloading medical facilities and causing fear in the population. Mustard gas is the best known blister agent.

blood agents—Agents based on cyanide compounds. More likely to be used for assassination than for terrorism.

botulism—An illness caused by the botulinum toxin, which is exceedingly lethal and quite simple to produce. It takes just a small amount of the toxin to destroy the central nervous system. Botulism may be contracted by the ingestion of contaminated food or through breaks or cuts in the skin. Food supply contamination or aerosol dissemination of the botulinum toxin are the two ways most likely to be used by terrorists.

Bush Doctrine—The policy that holds responsible nations which harbor or support terrorist organizations and says that such countries are considered hostile to the U.S. From President Bush's speech: "A country that harbors terrorists will either deliver the terrorist or share in their fate. . . . People have to choose sides. They are either with the terrorists, or they're with us."

business continuity—The ability of an organization to continue to function before, during, and after a disaster.

BWC—Officially known as the "Convention on the Prohibition of Development, Production, and Stockpiling of Bacteriological (Biological) and Toxic Weapons and Destruction." The BWC works toward general and complete disarmament, including the prohibition and elimination of all types of weapons of mass destruction.

Camp X-Ray—Guantanamo Bay, Cuba, which houses al-Qaeda and Taliban prisoners.

carrier—A person or animal that is potentially a source of infection by carrying an infectious agent without visible symptoms of the disease.

cascading event—The occurrence of one event that causes another event.

causative agent—The pathogen, chemical, or other substance that is the cause of disease or death in an individual.

cell—The smallest unit within a guerrilla or terrorist group. A cell generally consists of two to five people dedicated to a terrorist cause. The formation of cells is born of the concept that an apparent "leaderless resistance" makes it hard for counterterrorists to penetrate.

chain of custody—The tracking and documentation of physical control of evidence.

chemical agent—A toxic substance intended to be used for operations to debilitate, immobilize, or kill military or civilian personnel.

chemical ammunition—A munition, commonly a missile, bomb, rocket, or artillery shell, designed to deliver chemical agents.

chemical attack—The intentional release of toxic liquid, gas or solid in order to poison the environment or people.

chemical warfare—The use of toxic chemicals as weapons, not including herbicides used to defoliate battlegrounds or riot control agents such as gas or mace.

chemical weapons—Weapons that produce effects on living targets via toxic chemical properties. Examples would be sarin, VX nerve gas, or mustard gas.

chemterrorism—The use of chemical agents in a terrorist operation. Well-known chemical agents include sarin and VX nerve gas.

choking agent—Compounds that injure primarily in the respiratory tract (i.e., nose, throat, and lungs). In extreme cases membranes swell up, lungs become filled with liquid, and death results from lack of oxygen.

Cipro—A Bayer antibiotic that combats inhalation of anthrax.

communications architecture elements—Assets, systems, and networks that make up the communications architecture. The following are sample categories of architecture elements:

- *core network/Internet backbone*—The portion of the communications network that consists of high-capacity network elements servicing nationwide, regional, and international connectivity.
- *signaling and control systems*—Systems that exchange information regarding the establishment of a connection and control the management of the network.
- *shared assets and systems*—Assets and systems owned and operated by multiple companies, including facilities where equipment is collocated and systems are shared by network operators.

- *access*—Primarily the local portion of the network connecting end users to the backbone that enables users to send or receive communications. Access includes equipment and systems such as Public Switched Telephone Network (PSTN) switches, asynchronous transfer mode (ATM) switches, video servers for video on demand, and Internet Protocol (IP) routers for Internet Service Providers (ISPs).
- *customer equipment*—Equipment owned and operated by the end user or located at the end user's facility. Customers include individuals, organizations, businesses, and government.

communications sector—Public and private sector entities that have equities in the provisioning, use, protection, or regulation of communications networks and services. The communications sector is made up of five industry sectors:

- *wireline.* Consists primarily of the PSTN, but also includes enterprise networks. The PSTN is a domestic communications network accessed by telephones, key telephone systems, private branch exchange (PBX) trunks, and data arrangements. Despite industry's transition to packet-based networks, the traditional PSTN remains the backbone of the communications infrastructure, including landline telephone, the Internet, and submarine cable infrastructure.
- *wireless.* Refers to telecommunication in which electromagnetic waves (rather than some form of wire) carry the signal over part of or the entire communication path. Consists of cellular phone, paging, personal communication services, high-frequency radio, unlicensed wireless, and other commercial and private radio services.
- *satellite.* This is a space vehicle launched into orbit to relay audio, data, or video signals as part of a telecommunications network. Signals are transmitted to the satellite from earth station antennas, amplified, and sent back to earth for reception by other earth station antennas. Satellites are capable of linking two points, one point with many others, or multiple locations with other multiple locations. Uses a combination of terrestrial and space components to deliver various communications, Internet data, and video services.

- *cable*. This wireline network offers television Internet, and voice services that interconnect with the PSTN through end offices. Primary cable television (CATV) network components include headends and fiber optic and/or HFC. Because the CATV network was designed primarily for downstream transmission of television signals, most of the existing network is being refitted to support two-way data transmissions.
- *broadcasting*. Broadcasting systems consist of free, over-the-air radio and television stations that offer analog and digital audio and video programming services and data services. Broadcasting has been the principal means of providing emergency alerting services to the public for six decades. Broadcasting systems operate in four frequency bands: medium frequency (MF (AM radio)), very high frequency (VHF (FM radio and television)), and ultra-high frequency (VHF (FM radio and television)), and ultra-high frequency (UHF (television)). The recent transition to digital television (DTV) and ongoing transition to digital radio provide broadcast stations with enhanced capabilities, include the ability to multicast multiple programs on a single channel.

"confirmed"—In the context of the threat evaluation process, a water contamination incident is definitive evidence that the water has been contaminated.
consequence—The effect of an event, incident, or occurrence. For the purposes of the NIPP, consequences are divided into four main categories: public health and safety, economic, psychological, and governance impacts.
control center—A sophisticated monitoring and control system responsible for balancing power generation and demand, monitoring flows over transmission lines to avoid overloading, planning and configuring the system to operate reliably, maintaining system stability, preparing for emergencies, and placing equipment in and out of service for maintenance and emergencies.
control systems—Computer-based systems used within many infrastructures and industries to monitor and control sensitive processes and physical functions. These systems typically collect measurement and operation data from the field, process and display the information, and relay control commands to local or remote equipment or human-machine interfaces (operators). Examples of types of control systems include SCADA systems, Process Control Systems, and Distributed Control Systems.

counterterrorism—Measures used to prevent, preempt, or retaliate against terrorist attacks.

"credible"—In the context of the threat evaluation process, a water contamination threat is characterized as "credible" if information collected during the threat evaluation process corroborates information from the threat warning.

critical infrastructure—Systems and assets, whether physical or virtual, so vital that the incapacity or destruction of such may have a debilitating impact on the security, economy, public health or safety, environment, or any combination of these matters, across any federal, state, regional, territorial, or local jurisdiction.

cutaneous—Related to or entering through the skin.

cutaneous anthrax—Anthrax that is contracted via broken skin. The infection spreads through the bloodstream, causing cyanosis, shock, sweating, and finally death.

cyanide agent—Used by Iraq in the Iran war against the Kurds in the 1980s, and also by the Nazis in the gas chambers of concentration camps, cyanide is a colorless liquid which is inhaled in its gaseous form while liquid cyanide and cyanide salts are absorbed by the skin. Symptoms are headache, palpitations, dizziness, and respiratory problems followed later by vomiting, convulsions, respiratory failure, unconsciousness, and eventually death.

cybersecurity—The prevention of damage to, unauthorized use of, or exploitation of, and, if needed, the restoration of electronic information and communications systems and the information contained therein to ensure confidentially, integrity, and availability. Cybersecurity includes protection and restoration, when needed, of information networks and wireline, wireless, satellite, public safety answering points, and 911 communications systems and control systems.

cyber system—Any combination of facilities, equipment, personnel, procedures, and communications integrated to provide cyber services. Examples include business systems, control systems, and access control systems.

cyberterrorism—Attacks on computer networks or systems, generally by hackers working with or for terrorist groups. Some forms of cyberterrorism include denial of service attacks, inserting viruses or stealing data.

Defense Industrial Base Sector—The DoD, U.S. Government (USG), and private sector worldwide industrial complex with capabilities to perform research, development, and design and to produce and maintain military weapon systems, subsystems, components, or parts to meet military requirements.

defense infrastructure sector—A virtual association with the defense critical infrastructure that traverses normal organization boundaries and encompasses defense networks, assets, and associated dependencies that perform similar functions within the Department of Defense and are essential to the execution of the National Defense Strategy.

defensive critical asset—An asset of such extraordinary importance to operations in peace, crisis, and war that its incapacitation or destruction would have a very serious, debilitating effect on the ability of the Department of Defense to fulfill its missions.

dependency—The one-directional reliance of an asset, system network, or collection thereof, within or across sectors, in input, interaction, or other requirement from other sources in order to function properly.

dirty bomb—A makeshift nuclear device which is created from radioactive nuclear waste material. While not a nuclear blast, an explosion of a dirty bomb causes localized radioactive contamination as the nuclear waste material is carried into the atmosphere where it is dispersed by the wind.

ebola—Ebola hemorrhagic fever (Ebola EF) is a severe, often fatal disease in nonhuman primates such as monkeys, chimpanzees, gorillas and in humans. Ebola has appeared sporadically since 1976 when it was first recognized.

eBomb (or e-bomb)—Electromagnetic bomb which produces a brief pulse of energy which affects electronic circuitry. At low levels, the pulse temporarily disables electronics systems, including computers, radios, and transportation systems. High levels completely destroy circuitry, causing mass disruption of infrastructure while sparing life and property.

Ecotage—The portmanteau of the "eco-" prefix and "sabotage." It is used to describe illegal acts of vandalism and violence committed in the name of environmental protection.

Ecoterrorism—A neologism for terrorism that includes sabotage intended to hinder activities that are considered damaging to the environment.

Electromagnetic Pulse (EMP)—A burst of electromagnetic radiation by deliberate means, such as nuclear attack, or through natural means, such as a large-scale geomagnetic storm. Magnetic and electric fields resulting from EMP have the potential to disrupt electrical and electronic systems by causing destructive current and voltage surges.

Energy Asset and System Parameters—Six general asset or system characteristics that are important parameters for evaluating the vulnerabilities of energy infrastructure and developing risk management programs. They include physical and location attributes, cyber attributes, volumetric or throughput attributes, temporal/load profile attributes, human attributes, and the importance of an asset or system to the energy network.

Euroterrorism—Associated with left-wing terrorism of the 1960s, 1970s, and 1980s involving the Red Brigade, Red Army Faction, and November 17th Group, among other groups which targeted American interests in Europe and NATO. Other groups include Orange Volunteers, Red Hand Defenders, Continuity IRA, Loyalist Volunteer Force, Ulster Defense Association, and First of October Anti-Fascist Resistance Group.

fallout—The descent to the earth's surface of particles contaminated with radioactive material from a radioactive cloud. The term can also be applied to the contaminated particulate matter itself.

Fatah—Meaning "conquest by means of jihad"; a political organization created in the 1960s and led by Yasser Arafat. With both a military and intelligence wing, it has carried out terrorist attacks on Israel since 1965. It joined the PLO in 1968. Since 9/11, the Fatah was blamed for attempting to smuggle 50 tons of weapons into Israel.

fatwa—A legal ruling regarding Islamic Law.

Fedayeen Saddam—Iraq's paramilitary organization said to be equivalent to the Nazis' "SS". The militia is loyal to Saddam Hussein and is responsible for using brutality on civilians who are not loyal to the policies of Saddam. They do not dress in uniform.

filtrate—In ultrafiltration, the water that passes through the membrane and which contains particles smaller than the molecular weight cutoff of the membrane.

frustration-aggression hypothesis—A hypothesis that every frustration leads to some form of aggression and every aggressive act results from some prior frustration. As defined by Gurr: "The necessary precondition for violent civil conflict is relative deprivation, defined as actors' perception of discrepancy between their value expectations and their environment's apparent value capabilities. This deprivation may be individual or collective."

function—A service, process, capability, or operation performed by an asset, system, network, or organization.

fundamentalism—Conservative religious authoritarianism. Fundamentalism is not specific to Islam; it exists in all faiths. Characteristics include literal interpretation of scriptures and a strict adherence to traditional doctrines and practices.

Geneva Protocol 1925—The first treaty to prohibit the use of biological weapons. The 1925 Geneva Protocol for the Prohibition of the Use in War of Asphyxiating, Poisonous or Other Gases and Bacteriological Methods of Warfare.

germ warfare—The use of biological agents to cause harm to targeted people either directly, by bringing the people into contact with the agents, or indirectly, by infecting other animals and plants, which would in turn cause harm to the people.

glanders—An infectious bacterial disease known to cause inflammation in horses, donkeys, mules, goats, dogs and cats. Human infection has not been seen since 1945, but because so few organisms are required to cause disease, it is considered a potential agent for biological warfare.

Government Coordinating Council (GCC)—The government counterpart to the SCC of each sector established to enable interagency coordination. The GCC comprises representatives across various levels of government (federal, state, local, tribal, and territorial) as appropriate to the security and operations landscape of each individual sector.

grab sample—A single sample collected at a particular time and place that represents the composition of the water, air, or soil only at that time and location.

ground zero—From 1946 until 9/11, ground zero was the point directly above, below, or at which a nuclear explosion occurs or the center or origin of rapid, intense, or violent activity or change. After 9/11, the term, when used with initial capital letters, refers to the ground at the epicenter of the World Trade Center attacks.

guerrilla warfare—The term was invented to describe the tactics Spain used to resist Napoleon, though the tactic itself has been around much longer. Literally, it means "little war." Guerilla warfare features cells and utilizes no front line. The oldest form of asymmetric warfare, guerilla warfare is based on sabotage and ambush with the objective of destabilizing the government through lengthy and low-intensity confrontation.

Hamas—A radical Islamic organization which operates primarily in the West Bank and Gaza Strip whose goal is to establish an Islamic Palestinian state in place of Israel. On the one hand, Hamas operates overtly in their capacity as social services deliverers, but its activists have also conducted many attacks, including suicide bombings, against Israeli civilians and military targets.

hazard—An inherent physical or chemical characteristic that has the potential for causing harm to people, the environment, or property.

hazard assessment—The process of evaluating available information about the site to identify potential hazards that might pose a risk to the site characterization team. The hazard assessment results in assigning one of four levels to risk: lower hazard, radiological hazard, high chemical hazard, or high biological hazard.

hemorrhagic fevers—In general, the term viral hemorrhagic fever is used to describe a severe multisystem syndrome wherein the overall vascular system is damaged and the body becomes unable to regulate itself. These symptoms are often accompanied by hemorrhage; however, the bleeding itself is not usually life-threatening. Some types of hemorrhagic fever viruses can cause relatively mild illnesses.

High-Impact, Low-Frequency (HILF)—HILF events are occurrences that are relatively unusual, but have the potential to cause catastrophic disruption. Examples include pandemic disease, terrorist attack, and electromagnet pulse.

Hizbollah (Hezbollah)—Meaning "The Party of God." One of many terrorist organizations which seek the destruction of Israel and the United States. They have taken credit for numerous bombings against civilians, and have declared that civilian targets are warranted. Hezbollah claims it sees no legitimacy for the existence of Israel, and that their conflict becomes one of legitimacy that is based on religious ideals.

Homeland Security Office—An agency organized after 9/11, with former Pennsylvania Governor Tom Ridge heading it up. The Office of Homeland Security is at the top of approximately 40 federal agencies charged with protecting the U.S. against terrorism.

homicide bombings—A term the White House coined to replace the old "suicide bombings."

incident—A confirmed occurrence that requires response actions to prevent or minimize loss of life or damage to property and/or natural resources. A drinking water contamination incident occurs when the presence of a harmful contaminant has been confirmed.

infrastructure—The framework of interdependent networks and systems comprising identifiable industries, (including people and procedures), and distribution capabilities that provide a reliable flow of products and services essential to the defense of economic security of the United States, the smooth functioning of government at all levels, and society as a whole. Consistent with the definition in the Homeland Security Act, infrastructure includes physical, cyber, and/or human elements.

inhalation anthrax—A form of anthrax that is contracted by inhaling anthrax spores. This results in pneumonia, sometimes meningitis, and finally death.

interdependency—Mutually reliant relationship between entities (objects, individuals, or groups). The degree of interdependency does not need to be equal.

intifada (intifadah)—(alternatively Intifadah, from Arabic "shaking off") The two intifadas are similar in that both were originally characterized by civil disobedience by the Palestinians which escalated into the use of terror. In 1987, following the killing of several Arabs in the Gaza Strip, the first intifada began and went on until 1993. The second intifada began in September 2000, following Ariel Sharon's visit to the Temple Mount.

ISIS (ISIL) Islamic State of Iraq and the Levant—Islamic extremist rebel group that presently controls territory and brutalizes and/or murders various groups in various countries, including Syria, Iraq, Libya, Nigeria, and others.

Islam—Meaning "submit." The faith practiced by followers of Muhammad. Islam claims more than a billion believers worldwide.

jihad—Meaning "struggle." The definition is a subject of vast debate. There are two definitions generally accepted. The first is a struggle against oppression, whether political or religious. The second is the struggle within oneself, or a spiritual struggle.

key resources—As defined in the Homeland Security Act, key resources are publicly or privately controlled resources essential to the minimal operations of the economy and government.

kneecapping—A malicious wounding by firearm to damage the knee joint; a common punishment used by Northern Ireland's IRA on those involved in collaborating with the British.

Koran—The holy book of Islam, considered by Muslims to contain the revelations of God to Mohammed. Also called Qu'ran.

Laboratory Response Network (LRN)—A network of labs developed by the CDC, APHL, and FBI for the express purpose of dealing with bioterrorism threats, including pathogens and some biotoxins.

lassa fever—An acute, often fatal, viral disease characterized by high fever, ulcers of the mucous membranes, headaches, and disturbances of the gastrointestinal system.

LD50—The dose of a substance which kills 50 percent of those infected.

links—The means (road, rail, barge, or pipeline) by which a chemical is transported form one node to another.

mindset—According to *American Heritage Dictionary*: "1. A fixed mental attitude or disposition that predetermines a person's response to an interpretation of situations; 2. an inclination or a habit." *Merriam Webster's Collegiate Dictionary* (10th ed.) defines it as 1. A mental attitude or inclination; 2. a fixed state of mind. The term dates from 1926 but apparently is not included in dictionaries of psychology.

mitigation—Ongoing and sustained action to reduce the probability of or lessen the impact of an adverse incident.

Molotov cocktail—A crude incendiary bomb made of a bottle filled with flammable liquid and fitted with a rag wick.

monkeypox—The Russian bioweapon program worked with this virus, which is in the same family as smallpox. In June 2003, a spate of human monkeypox cases was reported in the U.S. Midwest. This was the first time that monkeypox was seen in North America, and it was the first time that monkeypox was transferred from animal to human. There was some speculation that it was a bioattack.

mullah—A Muslim, usually holding an official post, who is trained in traditional religious doctrine and law.

Muslim (also Moslem)—Followers of the teachings of Mohammed, or Islam.

mustard gas—Blistering agents which cause severe damage to the eyes, internal organs, and respiratory system. Produced for the first time in 1822, mustard gas was not used until World War I. Victims suffered the effects of mustard gas 30 to 40 years after exposure.

narcoterrorism—The view of many counterterrorist experts that there exists an alliance between drug traffickers and political terrorists.

National Pharmaceutical Stockpile—A stock of vaccines and antidotes which are stored at the Centers for Disease control in Atlanta, to be used against biological warfare.

nerve agent—The Nazis used the first nerve agents: insecticides developed into chemical weapons. Some of the better known nerve agents include VX, sarin, soman, and tabun. These agents are used because only a small quantity is necessary to inflict substantial damage. Nerve agents can be inhaled or can absorb through intact skin.

network—A group of components that share information or interact with each other in order to perform a function.

node—A facility at which a chemical is produced, stored, or consumed.

nuclear blast—An explosion of any nuclear material which is accompanied by a pressure wave, intense light and heat, and widespread radioactive fallout which can contaminate the air, water and ground surface for miles around.

opportunity contaminant—A contaminant that might be readily available in a particular area, even though it may not be highly toxic or infectious or easily dispersed and stable in treated drinking water.

Osama bin Laden (also spelled "Usama")—A native of Saudi Arabia, was born the 17th of 24 sons of Saudi Arabian builder Mohammed bin Oud bin Laden, a Yemeni immigrant. Early in his career, he helped the mujahedeen fight the Soviet Union by recruiting Arabs and building facilities. He hates the United States and apparently this is because he views the United States as having desecrated holy ground in Saudi Arabia with their presence during the first Gulf War. Expelled from Saudi Arabia in 1991 and from Sudan in 1996, he operated terrorist training camps in Afghanistan. His global network, al-Qaeda, is credited with the attacks on the United States on September 11, 2001, the attack on the USS Cole in 2000, and a number of other terrorist attacks.

owners/operators—Those entities responsible for day-to-day operation and investment in a particular asset or system.

pandemic influenza—Defined by the World Health Organization (WHO) as a global outbreak of influenza, characterized by an emergent strain of the virus, little to no immunity among the general population, rapid and sustained person-to-person transmission, and lack of a vaccine. On June 11, 2009, WHO determined that 2009 H1N1 influenza (also known as "swine flu") had reached pandemic status.

pathogen—Any agent which can cause disease.

pathways—The sequence of nodes and links by which a chemical is produced, transported, and transformed from its initial source to its ultimate consumer.

physical security—The use of barriers and surveillance to protect resources, personnel, and facilities against crime, damage, or unauthorized access.

plague—The pneumonic plague, which is more likely to be used in connection with terrorism, is naturally carried by rodents and fleas but can be aerosolized and sprayed from crop dusters. A 1970 World Health Organization assessment asserted that, in a worst-case scenario, a dissemination of 50 kg in an aerosol over a city of 5 million could result in 150,000 cases of pneumonic plague, 80,000–100,000 of which would require hospitalization, and 36,000 of which would be expected to die.

political terrorism—Terrorist acts directed at governments and their agents and motivated by political goals (i.e., national liberation).

"possible"—In the context of the threat evaluation process, a water contamination threat is characterized as "possible" if the circumstances of the threat warning appear to have provided an opportunity for contamination.

potassium iodide—An FDA-approved nonprescription drug for use as a blocking agent to prevent the thyroid gland from absorbing radioactive iodine.

preparedness—The activities necessary to build, sustain, and improve readiness capabilities to prevent, protect against, respond to, and recover from natural or manmade incidents. Preparedness is a continuous process involving efforts at all levels of government and between government and the private sector and nongovernmental organizations to identify threats, determine vulnerabilities, and identify required resources to prevent, respond to, and recover from major incidents.

presumptive results—Results of chemical and/or biological field testing that need to be confirmed by further lab analysis. Typically used in reference to the analysis of pathogens.

prevention—Actions taken and measures put in place for the continual assessment and readiness of necessary actions to reduce the risk of threats and vulnerabilities, to intervene and stop an occurrence, or to mitigate effects.

prioritization—In the context of the NIPP, prioritization is the process of using risk assessment results to identify where risk reduction or mitigation efforts are most needed and to subsequently determine which protective actions should be instituted in order to have the greatest effect.

protection—Actions or measures taken to cover or shield from exposure, injury, or destruction. In the context of the NIPP, protection includes actions to deter the threat, mitigate the vulnerabilities, or minimize the consequences associated with a terrorist attack or other incident. Protection can include a wide range of activities, such as hardening facilities, building resilience and redundancy, incorporating hazard resistance into initial facility design, initiating active or passive countermeasures, installing security systems and redundancy, incorporating hazard resistance into initial facility design, initiating active or passive countermeasures, installing security systems, promoting workforce surety, training and exercises, and implementing cybersecurity measures, among various others.

psychopath—A mentally ill or unstable person, especially one having a psychopathic personality (*q.v.*), according to *Webster's*.

psychopathology—The study of psychological and behavioral dysfunction occurring in mental disorder or in social disorganization, according to *Webster's*.

psychopathy—A mental disorder, especially an extreme mental disorder marked usually by egocentric and antisocial activity, according to *Webster's*.

psychotic—Of, relating to, or affected with psychosis, which is a fundamental mental derangement (as schizophrenia) characterized by defective or lost contact with reality, according to *Webster's*.

rapid field testing—Analysis of water during site characterization uses rapid field water testing technology in an attempt to tentatively identify contaminants or unusual water quality.

recovery—The development, coordination, and execution of service- and site-restoration plans for affected communities and the reconstitution of government operations and services through individual, private sector, nongovernmental, and public assistance programs that identify needs and define resources; provide housing and promote restoration; address long-term care and treatment of affected persons; implement additional measures for community restoration; incorporate mitigation measures and techniques, as flexible; evaluate the incident to identify lessons learned and develop initiatives to mitigate the effects of future incidents.

red teaming—As used in this text, a group exercise to imagine all possible terrorist attack scenarios against the chemical infrastructure and their consequences.

redundancy—An energy reliability strategy based on the notion that multiple systems provide needed backup if one system fails or cannot meet demand.

resilience/resiliency—The ability to resist, absorb, recover from, or successfully adapt to adversity or a change in conditions. In the context of energy security, resilience is measured in terms of robustness, resourcefulness, and rapid recovery.

response—Activities that address the short-term, direct effects of an incident, including immediate actions to save lives, protect property, and meet basic human needs. Response also includes the execution of emergency operations plans and incident mitigation activities designed to limit the loss of life, personal injury, property damage, and other unfavorable outcomes. As indicated by the situation, response activities include applying intelligence and other information to lessen the effects or consequences of an incident; increasing security operations; continuing investigations into the nature and source of the threat; ongoing surveillance and testing processes; immunizations, isolation, or quarantine; and specific law enforcement operations aimed at preempting, interdicting, or disrupting illegal activity, and apprehending actual perpetrators and bringing them to justice.

retentate—In ultrafiltration, the retentate is the solution that contains the particles that do not pass through the membrane filter. The retentate is also called the concentrate.

ricin—A stable toxin easily made from the mash that remains after processed castor beans. At one time, it was used as an oral laxative, castor oil; castor oil causes diarrhea, nausea, vomiting, abdominal cramps, internal bleeding, liver and kidney failure, and circulatory failure. There is no antidote.

risk—The potential for an unwanted outcome resulting from an incident, event, or occurrence, as determined by its likelihood and the associated consequences.

salmonella—An infection caused by a gram-negative bacillus, a germ of the Salmonella genus. Infection with this bacteria may involve only the intestinal tract or may be spread from the intestines to the bloodstream and then to other sites in the body. Symptoms of salmonella enteritis include diarrhea, nausea, fever, and abdominal pain. Dehydration resulting from the diarrhea can cause death, and the disease could cause meningitis or septicemia. The incubation period is between 8 and 48 hours, while the acute period of the illness can hang on for one to two weeks.

sarin—A colorless, odorless gas. With a lethal dose of .5 mg (a pinprick-sized droplet), it is 26 times more deadly than cyanide gas. Because the vapor is heavier than air, it hovers close to the ground. Sarin degrades quickly in humid weather, but sarin's life expectancy increases as temperatures get higher, regardless of how humid it is.

sector—A logical collection of assets, systems, or networks that provide a common function to the economy, government, or society. The NIPP addresses eighteen Critical Infrastructures sectors, identified by the criteria set forth in HSPD-7.

Sentinel Laboratory—An LRN lab that reports unusual results that might indicate a possible outbreak, and refers specimens that may contain select biological agents in reference labs within the LRN.

site characterization—The process of collecting information from an investigation site in order to support the evaluation of a drinking water contamination threat. Site characterization activities include the site investigation, field safety screening, rapid field testing of the water, and sample collection.

situational awareness—An understanding of the current environment and the ability to accurately anticipate future problems in order to respond effectively.

sleeper cell—A small cell which keeps itself undetected until such time as they can "awaken" and cause havoc.

smallpox—The first biological weapon, used during the eighteenth century, smallpox killed 300 million people in the nineteenth century. There is no specific treatment for smallpox disease, and the only prevention is vaccination. This currently poses a problem, since the vaccine was discontinued in 1970 and the WHO declared smallpox eradicated. Incubation is seven to seventeen days, during which the carrier is not contagious. Thirty percent of people exposed are infected, and it has a 30 percent mortality rate.

smart grid—The electric delivery network, from electrical generation to end-use customer, integrated with the latest advances in digital and information technology to improve customer service and integrate with the latest advances in digital and information technology to improve electric-system reliability, security, and efficiency.

sociopath—Basically synonymous with psychopath (q.v.). Sociopathic symptoms in the adult sociopath include an inability to tolerate delay or frustration, a lack of feeling guilty, a relative lack of anxiety, a lack of compassion for others, a hypersensitivity to personal ills, and a lack of responsibility. Many authors prefer the term sociopath because this type of person had defective socialization and a deficient childhood.

sociopathic—Of, relating to, or characterized by asocial or antisocial behavior or a psychopathic (q.v.) personality, according to *Webster's*.

spore—An asexual, usually single-celled reproductive body of plants such as fungi, mosses or ferns; a microorganism, as a bacterium, in a resting or dormant state.

system—Any combination of facilities, equipment, personnel, procedures, and communications integrated for a specific purpose.

terrorism—Premeditated threat or act of violence against non-combatant persons, property, and environmental or economic targets to induce fear, intimidate, coerce, or affect a government, the civilian population, or any segment thereof, in furtherance of political, social, ideological, or religious objectives.

terrorist group—A group which practices or has significant elements which are involved in terrorism.

threat—An indication that a harmful incident, such as contamination of the drinking water supply, may have occurred. The threat may be direct, such as a verbal or written threat, or circumstantial, such as a security breach or unusual water quality.

toxin—A poisonous substance produced by living organisms capable of causing disease when introduced into the body tissues.

transponder—A device on an airliner which sends out a signal allowing air traffic controllers to track an airplane. Transponders were disabled in some of the planes high-jacked on 9/11.

Transportation Security Administration (TSA)—A new agency created by the Patriot Act of 2001 for the purpose of overseeing technology and security in American airports.

tularemia—An infectious disease caused by a hardy bacterium, *Francisella tularensis*, found in animals, particularly rabbits, hares, and rodents. Symptoms depend upon how the person was exposed to tularemia but can include difficulty breathing, chest pain, bloody sputum, swollen and painful lymph glands, ulcers on the mouth or skin, swollen and painful eyes, and sore throat. Symptoms usually appear 3 to 5 days after exposure, but sometimes will take up to two weeks. Tularemia is not spread from person to person, so people who have it need not be isolated.

ultrafiltration—A filtration process for water that uses membranes to preferentially separate very small particles that are larger than the membrane's molecular weight cutoff, typically greater than 10,000 Daltons. (A Dalton is a unit of mass, defined as 1/12 the mass of a carbon-12 nucleus. It's also called the atomic mass unit, abbreviated as either "amu" or "u".)

value proposition—A statement that outlines the national and homeland security interest in protecting the nation's critical infrastructure and articulates the benefits gained by all critical infrastructure partners through the risk management framework and public-private partnership described in the NIPP.

vector—An organism which carries germs from one host to another.

vesicle—A blister filled with fluid.

vulnerability—A physical feature or operational attribute that renders an entity open to exploitation or susceptible to a given hazard.

weapons of mass destruction (WMD)—According to the National Defense Authorization Act, any weapon or device that is intended, or has the capability, to cause death or serious bodily injury to a significant number of people through the release, dissemination, or impact of:

- toxic or poisonous chemicals or their precursors
- a disease organism
- radiation or radioactivity

xenophobia—Irrational fear of strangers or those who are different from oneself.

zyklon b—A form of hydrogen cyanide. Symptoms of inhalation include increased respiratory rate, restlessness, headache, and giddiness, followed later by convulsions, vomiting, respiratory failure and unconsciousness. Used in the Nazi gas chambers in WWII.

Index

About the Author

Frank R. Spellman, PhD, is a retired assistant professor of environmental health at Old Dominion University, Norfolk, Virginia, and the author of more than 126 books covering topics ranging from concentrated animal feeding operations (CAFOs) to all areas of environmental science and occupational health. Many of his texts are readily available online, and several have been adopted for classroom use at major universities throughout the United States, Canada, Europe, and Russia; two have been translated into Spanish for South American markets. Dr. Spellman has been cited in more than 450 publications. He serves as a professional expert witness for three law groups and as an incident/accident investigator for the U.S. Department of Justice and a northern Virginia law firm. In addition, he consults on homeland security vulnerability assessments for critical infrastructures including water/wastewater facilities nationwide and conducts pre-Occupational Safety and Health Administration (OSHA)/Environmental Protection Agency (EPA) audits throughout the country. Dr. Spellman receives frequent requests to co-author with well-recognized experts in several scientific fields; for example, he is a contributing author of the prestigious text *The Engineering Handbook*, Second Edition. Dr. Spellman lectures on sewage treatment, water treatment, and homeland security and safety topics throughout the country and teaches water/wastewater operator short courses at Virginia Tech, Blacksburg, Virginia. He surveyed the drinking water ancient distribution system at Machu

Picchu and studied Darwin's finches in the Galapagos Islands. In addition, he studied water pollution in Coco, Ecuador. He holds a BA in public administration, a BS in business management, an MBA, and an MS and PhD in environmental engineering.

www.ingramcontent.com/pod-product-compliance
Lightning Source LLC
LaVergne TN
LVHW090314210926
848480LV00001B/242